What sort of times are these, when
To talk about trees is almost a crime,
Because it is simultaneously silence about so many atrocities!

Bertolt Brecht, *"An die Nachgeborenen,"* 1939

Talking About Trees

Talking About Trees

Science, Ecology and Agriculture in Cuba

Richard Levins

LeftWord

Offset edition first published in July 2008.
Reprinted as paperback in June 2011.
Digital print edition, January 2020.

LeftWord Books
2254/2A Shadi Khampur
New Ranjit Nagar
New Delhi 110008
INDIA

www.leftword.com

LeftWord Books is a division of
Naya Rasta Publishers Pvt. Ltd.

ISBN 978-81-87496-63-2 HC
978-81-87496-76-2 PB

GERARDO HERNANDEZ

RENE GONZALEZ

RAMON LABANINO

ANTONIO GUERRERO

FERNANDO GONZALEZ

Cubans condemned to long sentences by a Miami court for combating U.S.-based terrorism against Cuba. From their prison cells they continue steadfast in the struggle for justice.

Contents

Preface
Living the 11th Thesis

> Philosophers have sought to understand the world. The point, however, is to change it.
>
> Karl Marx (Theses on Feuerbach, 11th Thesis)

When I was a boy I always assumed that I would grow up to be both a scientist and a Red. Rather than face a problem of combining activism and scholarship, I would have had a very difficult time trying to separate them.

Before I could read, my grandfather read to me from Bad Bishop Brown's Science and History for Girls and Boys. He believed that as a minimum every socialist worker should be familiar with cosmology, evolution, and history. I never separated history, in which we are active participants, from science, the finding out how things are. My family had broken with organized religion five generations back, but my father sat me down for Bible study every Friday evening because it was an important part of the surrounding culture and important to many people, a fascinating account of how ideas develop in changing conditions, and because every atheist should know it as well as believers do.

On my first day of primary school, my grandmother urged me to learn everything they could teach me—but not to believe it all. She was all too aware of the "racial science" of 1930s Germany and the justifications for eugenics and male supremacy that were popular in our own country. Her attitude came from her knowledge of the uses of science for power and profit and from a worker's generic distrust of the rulers. Her advice formed my stance in academic life: consciously in, but not of, the university.

I grew up in a left-wing neighbourhood of Brooklyn where the schools were empty on May Day and where I met my first Republican at age 12. Issues of science and politics and culture were debated in permanent clusters on the Brighton Beach boardwalk and were the bread and butter of mealtime conversation. Political commitment was assumed, how to act on that commitment was a matter of fierce debate.

As a teenager I became interested in genetics through my fascination with the work of the Soviet scientist Lysenko. He turned out to be dreadfully wrong especially in trying to reach biological conclusions from philosophical principles. However, his criticism of the genetics of his time turned me toward the work of C.H. Waddington and I.I. Schmalhausen (genetic biologists of the twentieth century) and others who would not simply dismiss him out of hand in Cold War fashion but had to respond to his challenge by developing a deeper view of the organism/environment interaction

My wife, Rosario Morales, introduced me to Puerto Rico in 1951 and my eleven years there gave a Latin American perspective to my politics. The recent various left-wing victories in South America are a source of optimism even in these grim times. FBI surveillance in Puerto Rico blocked me from the jobs I was looking for and I ended up doing vegetable farming for a living on the island's western mountains.

As an undergraduate at Cornell University's School of Agriculture, I had been taught that the prime agricultural problem of the United States was the disposal of the farm surplus. But as a farmer in a poor region of Puerto Rico, I saw the significance of agriculture for peoples' lives. That experience introduced me to the realities of poverty as it undermines health, shortens lives, closes options, and stultifies personal growth, and to the specific forms that sexism takes among the rural poor. Direct labour organizing on the coffee plantations was combined with study. Rosario and I wrote the agrarian programme of the Puerto Rican Communist Party in which we combined rather amateurish economic and social analysis with some first insights into ecological production methods, diversification, conservation, and cooperatives.

I first went to Cuba in 1964 to help develop their population genetics and get a look at the Cuban revolution. Over the years I became involved in the ongoing Cuban struggle for ecological agriculture and an ecological pathway of economic development that was just, egalitarian, and sustainable. Progressivist thinking, so powerful in the socialist tradition, expected that developing countries had to catch up with advanced countries along the single pathway of modernization. It dismissed critics of the high-tech pathway of industrial agriculture as "idealists", urban sentimentalists nostalgic for a bucolic rural golden age that never really existed. But there was another view, that each society creates its own ways of relating to the rest of nature, its own pattern of land use, its own appropriate technology, its own criteria of efficiency. This discussion raged in Cuba in the 1970s and by the 80s the ecological model had basically won although implementation was still a long process. The Special Period, that time of economic crisis after the collapse of the Soviet Union when the materials for high tech became unavailable, allowed ecologists by conviction to recruit the ecologists by necessity. This was possible only because the ecologists by conviction had prepared the way.

I first met dialectical materialism in my early teens through the writings of the British Marxist scientists J.B.S. Haldane, J.D.

Bernal, Joseph Needham, and others, and then on to Marx and Engels. It immediately grabbed me both intellectually and aesthetically. A dialectical view of nature and society has been a major theme of my research since. I have delighted in the dialectical emphasis on wholeness, connection and context, change, historicity, contradiction, irregularity, asymmetry, and the multiplicity of levels of phenomena, a refreshing counterweight to the prevalent reductionism then and now.

An example: after Rosario suggested I look at Drosophila (fruit fly) in nature not just in bottles in the laboratory I started to work with the Drosophila in the neighbourhood of our home in Puerto Rico. My question was: how do Drosophila species cope with the temporal and spatial gradients of their environments? I began examining the multiple ways that different Drosophila species responded to similar environmental challenges. I could collect Drosophila in a single day in the deserts of Gúanica and in the rain forest around our farm at the crest of the cordillera. It turned out that some species adapt physiologically to high temperature in 2-3 days, and show relatively little genetic differences in heat tolerance along a 3000-foot altitude gradient (about 20 miles). Others had distinct genetic subpopulations in the different habitats. Still others adapted to and inhabited only a part of the available environmental range. One of the desert species was not any better at tolerating heat than some Drosophila from the rain forest, but were much better at finding the cool moist microsites and hiding in them after about 8 AM. These findings led me to describe the concepts of co-gradient selection, where the direct impact of the environment enhances genetic differences among populations, and counter-gradient selection where genetic differences offset the direct impact of the environment. Since on my transect the high temperature was associated with dry conditions, natural selection acted to increase the size of the flies at Guánica while the effect of temperature on development made them smaller. The outcome turned out to be that the flies from the sea level desert and the rain forest were of about the same size in their own habitats but the Guánica flies were bigger when raised at the same temperature as rain forest flies.

In this work I questioned the prevailing reductionist bias in biology by insisting that phenomena take place on different levels, each with their own laws, but also connected.

My bias was dialectical: the interaction among adaptations on the physiological, behavioral, and genetic levels. My preference for process, variability, and change set the agenda for my thesis.

The problem was how species can adapt to an environment when the environment wasn't always the same. When I began thesis work I was puzzled by the facile assumption that, faced with opposing demands, for example when the environment favours small size some of the time and large size the rest of the time, an organism would have to adopt some intermediate state as a compromise. But this is an unthinking application of the liberal bromide that when there are opposing views the truth lies somewhere in the middle. In my dissertation, the study of fitness sets was an attempt to examine when an intermediate position is truly an optimum and when is it the worst possible choice. The short answer turned out to be that when the alternatives are not too different, an intermediate position is indeed optimal but when they are very different compared to the range of tolerance of the species, then one extreme alone or in some cases a mixture of extremes is preferable.

Work in natural selection within population genetics almost always assumed a constant environment, but I was interested in its inconstancy. I proposed that "environmental variation" must be an answer to many questions of evolutionary ecology and that organisms adapt not only to specific environmental features such as high temperature or alkaline soils but also to the pattern of the environment—its variability, its uncertainty, the grain of its patchiness, the correlations among different aspects of the environment. Moreover, these patterns of environment are not simply given, external to the organism: organisms select, transform, and define their own environments.

Regardless of the particular matter of an investigation (evolutionary ecology, agriculture, or more recently, public health), my core interest has always been the understanding of the dynamics of complex systems. Also, my political commitment requires that I question the relevance of my work. In one of Brecht's poems he says "What sort of times are these, when to talk about trees is almost a crime, because it is simultaneously silence about so many atrocities!" Brecht was of course wrong about trees: nowadays when we talk of trees we are not ignoring injustice. But he was also right: scholarship that is indifferent to human suffering is immoral.

Poverty and oppression costs years of life and health, shrinks the horizons and cuts off potential talents before they can flourish. My commitment to support the struggles of the poor and oppressed and my interest in variability combined to focus my attention on the physiological and social vulnerabilities of people.

I have been studying the body's capacity to restore itself after it is stressed by malnutrition, pollution, insecurity, and inadequate health care. Continual stress undermines the stabilizing mechanisms in the bodies of oppressed populations making them more vulnerable to anything that happens, to small differences in their environments. This shows up in increased variability in measures blood pressure, body mass index, and life expectancy as compared to more uniform results in comfortable populations.

In examining the effects of poverty, it is not enough to

examine the prevalence of separate diseases in different populations. While specific pathogens or pollutants may precipitate specific named diseases, social conditions create more diffuse vulnerability that links medically unrelated diseases. For instance, malnutrition, infection, or pollution can breach the protective barriers of the intestine. But once breached for any of these reasons it becomes a locus of invasion by pollutants, microbes, or allergens. Therefore nutritional problems, infectious diseases, stress, and toxicities and cause a great variety of seemingly unrelated diseases.

The prevailing notion since the 1960s had been that infectious disease would disappear with economic development. In the 1990s I helped form the Harvard Group on New and Resurgent Disease that rejected that idea. Our argument was partly ecological: the rapid adaptation of vectors to changing habitats: to deforestation, irrigation projects, and population displacement by war and famine. And the equally rapid adaptation of pathogens to pesticides and antibiotics. But we also criticized the physical, institutional, and intellectual isolation of medical research from plant pathology and veterinary studies which could have shown sooner the broad pattern of upsurge: malaria and cholera and AIDS but also African swine fever, feline leukemia, tristeza disease of citrus, and bean golden mosaic virus. We have to expect epidemiological changes with growing economic disparities and with changes in land use, economic development, human settlement, and demography. The faith in the efficacy of antibiotics, vaccines, and pesticides against plant, animal, and human pathogens is naïve in the light of adaptive evolution. And the developmentalist expectation that economic growth will lead the rest of the world to affluence and to the elimination of infectious disease is being proved wrong by events.

The resurgence of infectious disease is but one

manifestation of a more general crisis: the eco-social distress syndrome, a pervasive multilevel crisis of dysfunctional relations within our species and between it and the rest of nature. It includes in one network of actions and reactions, patterns of disease, relations of production and reproduction, demography, our depletion and wanton destruction of natural resources, changing land use and settlement, and planetary climate change. It is more profound than previous crises, reaching higher into the atmosphere, deeper into the earth, more widespread in space and more long lasting, penetrating more corners of our lives. It is both a generic crisis of the human species and a specific crisis of world capitalism. Therefore it is a primary concern of both my science and my politics.

The complexity of this whole world syndrome can be overwhelming and yet to evade the complexity by taking the system apart to treat the problems one at a time can produce disasters. The great failings of scientific technology have come from posing problems in too small a way. Agricultural scientists who proposed the Green Revolution without taking pest evolution and insect ecology into account, and therefore expecting pesticides would control pests, have been surprised that pest problems increased with spraying. Similarly, antibiotics create new pathogens, economic development creates hunger, and flood control promotes floods. Problems have to be solved in their rich complexity; the study of complexity itself becomes an urgent practical as well as theoretical problem.

These interests inform my political work: within the left, my task has been to argue that our relations with the rest of nature cannot be separated from a global struggle for human liberation, while within the ecology movement my task had been to challenge the "harmony of nature" idealism of early environmentalism and to insist on identifying the social relations that lead to the present dysfunction. On the other hand my politics have determined my scientific ethics. I believe that all theories are wrong which promote, justify, or tolerate injustice.

A leftist critique of the structure of intellectual life is a counterweight to the culture of the universities and foundations. The anti-war movement of the 60s and 70s took up the issues of the nature of the university as an organ of class rule and made the intellectual community itself an object of theoretical as well as practical interest. I also joined Science for the People, an organization that started with a research strike at MIT in 1967 as a protest against military research on campus. As a member I helped in the challenge to the Green Revolution and genetic determinism. Anti-war activism also took me to Vietnam to investigate war crimes (especially the use of defoliants) and from there to organizing Science for Vietnam. We denounced the use of Agent Orange (used as defoliant in the Vietnamese jungle) that was causing birth defects among Vietnamese peasants. Agent Orange was one of the worst uses of chemical herbicides.

The Puerto Rican independence movement gave me an anti-imperialist consciousness that serves me well in a university that promotes "structural reform" and other euphemisms for empire. My wife's sharp working class feminism is a running source of criticism of the pervasive elitism and sexism. Regular work with Cuba shows me vividly that there is an alternative to a competitive, individualistic, exploitative society.

Community organizations, especially in marginalized communities, and the women's health movement, raise issues that academia prefers to ignore: the mothers of Woburn noticing that too many of their children from the same small neighbourhood had leukemia. The hundreds of environmental justice groups that noted that toxic waste dumps were concentrated in black and Latino neighbourhoods. The Women's Community Cancer project and others who insist on the environmental causes of cancer and other diseases while the university laboratories are looking for guilty genes. Their initiatives help me maintain an alternative agenda for both theory and action.

Within the university I have a contradictory relationship with the institution and with colleagues, a combination of cooperation and conflict.

We may share a concern about health disparities and persistent poverty, but are in conflict about corporations funding research for patentable molecules and about government agencies such as the U.S. Agency for International Development promoting the goals of empire.

I never aspired to what is conventionally considered a "successful career" in academia. I do not find most of my personal validation through the formal reward and recognition system of the scientific community and I try not to share the common assumptions of my professional community. This gives me wide freedom of choice. Thus when I declined to join the National Academy of Science and received many supportive letters praising my courage or calling it a difficult decision, I could honestly say that it was not a hard decision, merely a political choice taken collectively by the Science for the People group in Chicago. We judged that it was more useful to take a public stand against the Academy's collaboration with the Vietnam-American war than to join the Academy and attempt to influence its actions from inside. Dick Lewontin had already tried that unsuccessfully and resigned along with Bruce Wallace.

Most of my research has objectives at two levels: the particular problem at hand and some major theoretical or polemical issue. The study of temperature adaptation in fruit flies was also an argument for multiple levels of causation. Niche theory was also a foray into the interpenetration of opposites (organism and environment). Biogeography was about multiple levels of ecological and evolutionary dynamics. Ecological pest management was also a claim for whole-system strategies. Work on new and resurgent infectious disease combined biology and sociology. We examined why the public health community was caught by surprise when infectious disease would not go away. It therefore was an exercise in the self-examination of science.

I have always enjoyed mathematics and see one of its tasks as making the obscure obvious. I regularly employ a sort of mid-level math in unconventional ways to promote understanding more than prediction. Much modeling now aims at precise equations giving precise prediction. This makes sense in engineering. In the field of policy, it makes sense to those who are the advisors to the rulers who imagine they have complete enough control of the world to be able to optimize their efforts and investments of resources. But those of us who are in the opposition have no such illusion. The best we can do is to decide where to push the system. For this, a qualitative mathematics is more useful. My work with signed digraphs ("loop analysis") is one such approach. Rejecting the opposition between qualitative and quantitative analysis and the notion that quantitative is superior to qualitative, I have mostly worked with those mathematical tools that assist conceptualization of complex phenomena.

Political activism of course attracts the attention of the agencies of repression. I have been fortunate in that regard, having experienced only relatively light repression. Others did not fare as well, with lost careers, years of imprisonment, violent attacks, intense harassment even of their families, and deportations. Some, mostly from the Puerto Rican, Afro-American, and Native American liberation movements, as well as the five Cuban anti-terrorists arrested in Florida, are still political prisoners.

Exploitation kills and hurts people. Racism and sexism destroy health and thwart lives. Studying the greed and brutality and smugness of late capitalism is painful and infuriating.

Sometimes I have to recite from Jonathan Swift's *Ballad in a Bad Temper*:

> Like the boatman on the Thames
> I row by and call them names.
> Like the ever-laughing sage
> In a jest I spend my rage
> But it must be understood
> I would hang them if I could.

For the most part scholarship and activism have given me an enjoyable and rewarding life, doing work I find intellectually exciting, socially useful, and with people I love.

One
Dialectical Science and the Cuban Pathway

The world is grappling with a terrible dilemma: on the one hand the world's peoples and most governments aspire to a rising standard of living. On the other hand, if that rising standard of living were taken to mean following the pathway of development and attaining the life style of Euro-North America, it would destroy the life support system of all of us. The vast majority of the world's population can neither be kept in poverty nor drawn into the consumerist way of life. The goals of justice and sustainability seem to be mutually exclusive. But when two equally urgent goals are incompatible, it is often the case that we have posed the problem too small. As long as development and progress are seen as taking place along a single pathway from less developed to more developed, there is no solution. But what if development is seen as a branching pathway, with choices all along the way? What if a rising standard of living is seen mostly as a rising quality of life within the constraints of sustainability? What if economic development is not a goal in itself but a means to enriching life and preserving nature, with emphasis on equity, health, education, culture, recreation and mutual caring within an environment that is sustainable, diverse, and people-friendly? That is the unique pathway that Cuba has embarked on.

Every kind of society has its own pattern of relations of people with each other, with the production of life's necessities and with the rest of nature. Every major change in social form has carried with it changed relations with animate and inanimate nature — with plants and animals, with pathogens, with soil, water and air. What we see in Cuba is the emergence of a socialist society, with socialist patterns of production, demography, settlement, human relations, and relations with our habitat. Agriculture is one aspect of this process.

These changed relations develop unevenly in the course of getting by, solving problems and thinking about problems on the run. Nobody could really sit back and ask what should we do to have a socialist landscape, or forest or crops or technology. And yet, that is what is happening. Cuba is creating a socialist ecology, a socialist pattern of relations between humans and the land, the insects, the fungi and tomatoes. Sometimes it seems as if the changes are random as ideas are tried, abandoned, tried again in new ways. Sometimes it seems as if innovations are merely emergency responses to the embargo and the collapse of the Soviet bloc. And the people most immersed in the process have had little time to stand back from their works and say, "look, everybody, we're creating a socialist eco-social system!"

But that is what is happening. And that is what is missed when visitors admire particular innovations and hope to adopt some of them in their own countries: Cuban agriculture is a socialist agriculture, created under difficult conditions to meet socialist goals.

What, then is "socialist" about Cuban agriculture?

Under socialism, ownership is in the hands of the "associated producers" either directly in cooperatives or indirectly through the state. Where management by the state proved too indirect, inflexible and not sufficiently democratic, new forms of ownership were adopted. Farmers' organizations and local government often lead the particular production programmes. The various forms of rural democracy are needed not only for making the general policy decisions of the enterprise, but also for participating in nation-wide decisions and mobilizing the collective intelligence of the farm workers in solving problems jointly with the scientists.

Under capitalism we have the maximum plunder of the countryside by the rulers of the cities. This is redressed under socialism in several ways:

- Priority investment in the countryside and small towns, with guaranteed equal educational and health service opportunities. Often the most advanced technologies go first to the isolated areas in the mountains, such as photovoltaic solar energy collectors for schools off the national grid;
- Recycling of farm waste within the farm and the return of urban waste to the farms, which requires that these be non-toxic;
- Recruitment of city people to move to the countryside or to work there periodically;
- Development of urban and suburban agriculture, and of rural industry;
- Research aimed at meeting socially determined needs.

First, decision-making power and ownership are in the hands of the "associated producers" either directly in the various forms of cooperatives or indirectly through the socialist state. There is no single form of land tenure that is required by socialist theory. The transfer of land management from state farms to cooperatives was no abandonment of socialism but reorganization within socialism to meet socialist goals better. What is precluded is privately owned agribusiness, hiring wage labour to extract maximum profit, selection of crops for their market value and of technologies that undermine the health of the farm workers and the future capacity to produce. It also precludes a profit-driven agricultural inputs industry. This changes the priorities for research, the criteria for investment and the decisions about what and how to plant.

The top priority is feeding the whole people and directing the products of agriculture where they are most needed. The principle that everybody eats meant that as long as milk was in short supply it went to children and was not turned into high-value derivatives such as butter and cheese. (I recall the pitifully thin pizzas of 1968 when drought reduced milk production). It meant that research was directed toward the crops that people consumed and often sold to them below the costs of production. The questions of land use, of the division of labour among rural, suburban and urban agriculture, are solved by considerations of social need including production and employment.

A second priority is economic solvency. But this is not the same as profit maximization. Land is devoted to export crops to the extent that foreign exchange is needed to support development and production and to supplement what the population needs. Nor can economic solvency be achieved at the expense of deterioration of the soil, loss of bio-diversity or contributing to global warming. One way of increasing the economic viability of agriculture is reducing the use of off-the-farm inputs. Composting, recycling within the farm, biological pest management and use of bio-fertilizers fit into this priority. An option that is not available is downsizing, creating unemployment in order to improve the bottom line. On the contrary, urban agriculture is encouraged in part because it creates employment at the rate of about 20 jobs per hectare. Not all farms are solvent although they are moving in that direction. While the older ANAP (National Association of Small Producers) cooperatives are stable economically, only about half the newer UBPC (Basic Units of Cooperative Production) are already profitable. But given that the new cooperatives are developing in a socialist context, solvency can be expected as the members learn to manage.

A third concern is protection of people's health. This is not an "externality" the way it is calculated under capitalism. Anything that affects the life of the population is "internal" to the society and therefore a concern. This is one of the reasons for adopting ecological methods of pest management.

The way research is done is determined also by socialist goals and socialist means. Science is not a commodity, directed toward the production of the most profitable commodities in a race for patents. Whereas under capitalism a company directs much research effort to denying that their product does any harm, Cuban research seeks out the consequences of a technology as it percolates through the environment: every consequence is important and there are no externalities. Institutions are able to work together to meet long term or urgent goals without hiding their findings as proprietary secrets. Therefore the intellectual resources of the country can be mobilized toward shared social goals. And the full access of women to positions of leadership in science doubles the brain pool available for research.

Of course this process is not easy. The prestige of Euro-North Americans weighs heavily on the scientists of a developing country and acts against independent innovation.

Decisions about pesticides, about specialization, about livestock technology and mechanization followed prolonged debate and are still taking place. In Cuba as in the United Sates there were debates about pesticides. But a debate about pesticides in Cuba is different from a debate about pesticides in the United States. In Cuba, as in the United States, the debate can be frustrating. We can find stubbornness, conservatism, ignorance and even stupidity in any country. But what is different is that in Cuba the debates were expressions only of differences of opinion and therefore reason can eventually prevail. In capitalist countries, debates about technology are often weapons in the conflict of interests. The makers of pesticides never ask what might be the best way of reducing pest damage while protecting soil and people but rather, what is the best way to turn oil into marketable commodities to sell to farmers, and they defend their products with a ferocity driven by the bottom line.

The ecological transformation of Cuban agriculture since the early 1990s is overwhelmingly complex, including changes in agrotechnology, land tenure and use, social organization of production and research, educational programmes and financial structures. From the point of view of the participants it comes about through solving problems within the context of socialist goals, goals such as stable food production in the face of natural and economic uncertainty, input reduction, preservation of productive capacity, conservation of biodiversity, protection of people, stable and healthful food supplies; recycling of residues between town and country and ruralization of cities; control by the associated producers without specifying the form; research for social ends.

The ecological transformation of Cuban agriculture takes place in the context of ecological development as a whole — Rio, Kyoto, Montreal protocol on ozone; programmes against desertification; reforestation; protection of shorelines; integration of education, research, economics, ecology.

We can expect the future of Cuban agriculture to be an ecological agriculture. A mosaic pattern of land use is developing in which each plot of land contributes direct products but also contributes to the other plots: forests give lumber, charcoal, honey and nuts, but also shelter many species of wildlife, provide refuge for the birds and bats that consume pests, regulate the flow of water (thus reducing the dependence on fossil fuels to pump irrigation water), serve as wind breaks and barriers to the spread of pests, and create special microclimates at their edges to the distance of about ten times the height of the trees. Pastures produce the usual livestock products but also retard erosion as compared to row crops, collect manure, offer a diversity of nectar sources for bees and for parasitic wasps that infect pests. Livestock graze on the weeds of fruit orchards, and so on. Farming would take place in the countryside but also in suburban and urban areas where it ameliorates the harshness of the city landscape, brings perishable foods closer to the consumers, provides jobs, and strengthens the neighbourhood sense of community.

All of this would mean a system of great complexity adapted separately to each locality by those who work there. Ecological agriculture demands the combination of physical with intellectual labour by farmers, both a general long-term socialist goal and a way of creating jobs that could keep the highly educated Cubans in the countryside.

Two
Whose Scientific Method?

Opponents of the precautionary principle frequently express two criticisms that have to do with science. They argue that the precautionary principle itself is not "scientific" because it calls for action without full proof that the substance or activity in question is really harmful. These critics call, instead, for objective evidence through the "scientific method." And then they explain this failing of the precautionary principle by accusing supporters of the principle of being "anti-science" or "anti-technology."

In this chapter I mainly address the first reproach. But I would like to address, briefly the charge that advocates of the principle are anti-science, conjuring visions of Luddite machine-smashers who hearken back to an imaginary golden age — nostalgia for the man with the hoe on the part of people who have never tilled the soil. Proponents of the status quo imagine that there is only one kind of progress, their version of it, and that the only alternative is stagnation. They contrast their modern pesticides to doing nothing about pests, their modern antibiotics to newts' eyes and holly, their plastic cities to living in caves. Yet progress does not proceed along a single line from backward to advanced. Rather, progress is a branching pathway with choices along the way. The conflict is about those choices.

The progress of science is not the smooth penetration of light into dark corners. "Technology" in contemporary polemics usually refers to "high technology," that is, methods based on electronics, biochemistry, and informatics. But technology is merely the ability to do something, whether it is controlling mosquitoes by spraying with synthetic molecules or by putting a tight fitting cover on the well. The conflict is not between science and anti-science but between different pathways for science and technology: between a commodified science-for-profit and a gentle science for humane goals; between the sciences of the smallest parts and the sciences of dynamic wholes.

This chapter discusses the social construction of scientific production and the pattern of strength and weakness to which this leads. I offer proposals for a more holistic, integral approach to understanding and addressing environmental issues.

WHAT IS THE SCIENTIFIC METHOD?

Working scientists know that there is no such thing as the scientific method. There are many scientific methods, depending on the nature of the problem and the stage of the investigation.

Certain scientific traditions have developed to avoid at least the most obvious kinds of errors.

- We know that if B follows A, it does not necessarily mean that A caused B. Therefore we carry out controlled experiments. We might treat two groups of patients identically except that one group receives a new drug and the other the standard treatment or a placebo. Then we can claim with some confidence that the new drug does or does not cure the disease. (But only with *some* confidence: if the two groups of patients were both badly undernourished it may be that neither treatment helps or that joining a study of any kind improves the patients' condition. The things that are held constant can also affect the results).
- Scientists learned early that the expectations of participants can influence the outcomes of experiments, so that it is

important that patients do not know whether they are receiving the new drug or not. Then they learned that the expectations of the researchers can also affect the outcome, and the double blind was invented in which the researchers in contact with the patients do not know which ones belong to which group.

• All sorts of factors might affect the outcome, so we have learned to match the study populations for age, sex, income and other characteristics of possible relevance to the study outcome.

• Two groups of similar animals might differ just by chance. Therefore we make use of replication and statistical analysis, mathematical models that ask "what if" questions and formal procedures of hypothesis testing.

• Social criteria have also been developed: scientific evidence must be public, repeatable, and subject to criticism. Ideas must be examined independently of their source. Investigators should know and reveal their biases and conflicts of interest.

Formal discussion of "the scientific method" usually focuses on the last stages of an investigation: hypothesis testing. But investigative methods vary considerably. In some sciences a single critical experiment, after it is repeated in other labs, can be decisive. In others, knowledge is gained through the accumulation of many different kinds of evidence.

Thus the case for global climate change is not only the direct observation of a temperature rise but also the retreat of glaciers, the melting of icecaps, the upward and northward expansion of plant and animal species, the decline of coral reefs, fossil climates revealed in ice cores, tree rings, geological deposits, pollen succession, and so on. All of this is reinforced by a theoretical basis for the climate change in the accumulation of greenhouse gasses. Each kind of evidence may be questioned for specific technical reasons, but the massive accumulation of evidence and the diversity of the arguments are overwhelming.

Similarly, the evidence for the environmental causation of our cancer epidemic is based on its historical association with the oil industry and use of pesticides, the geography of cancers related to specific industries or molecules, the history of the rise in cancer in relation to specific carcinogenic exposures (such as the increase in smoking in the last century), differences among generations of immigrants; population studies; and laboratory research.

The various sciences have their own methods and conventions. Some sciences, such as physics and chemistry, are experimental in the narrow sense that allows replication. In these fields, relatively few kinds of objects occur in vast numbers and can be manipulated many times. Astronomy and geology, in contrast, do not allow for direct experiments, but the large number of stars permits controlled comparisons, and geological processes can be duplicated in the laboratory even if geo-history cannot. Anthropology has no precise replication and the numbers of distinct peoples in the world is small compared to numbers of molecules or mosquitoes. Further, the anthropological observer is very much a part of the system studied. (Quantum theory and relativity both introduce the observer as part of the system studied. It is now recognized as relevant to all sciences.) Human physiology and psychology often make use of animals, which are studied as proxies for humans. They differ from people, of course, but it is hoped that they are similar enough to permit transfer of conclusions to humans.

FAILURES OF THE SCIENTIFIC METHOD

The development of scientific method is aimed at avoiding the kinds of errors we have learned to worry about. And indeed, it has been successful in catching ordinary sloppiness, dirty glassware, division by zero, wishful thinking, and the individual idiosyncratic biases of scientists or their economic stake in their findings. It has been less successful at recognizing the shared biases of a whole scientific community, the beliefs that are so much a part of the common sense of the community that they are not even recognized as biases. Therefore it is instructive to note the history of some major collective scientific errors and failures.

The doctrine of the epidemiological transition proposed that infectious diseases would decline in importance over time.[1] Yet since the mid-1980s, the per cent contribution of infectious diseases to total mortality has been increasing. This is true even in developed countries and even if AIDS is excluded. Advocates of this doctrine failed to take into account the broad epidemiological pattern of waves of surging and ebbing of diseases. They also failed to examine thoroughly the disease profiles of other species; the ecology of disease, including the lesson that any major change in land use, vegetation, climate, human settlement, economics or technology is potentially also a major change in our relations with vectors and pathogens; the rapid evolution of resistance to antibiotics and pesticides; and the vulnerability of a socially and economically stratified population. They assumed that the toxicological fact that a chemical could kill insects in a bottle implied the ecological fact that widespread use of pesticides would control populations of the vectors that spread disease.

In agricultural science, the belief that mechanization, monoculture, and widespread use of chemicals would allow increased production to eliminate hunger failed to consider how pesticides exacerbate pest problems by disrupting the communities of species and provoking resistance; that monoculture increases vulnerability to invasion by pests and to unexpected climatic events; and that agribusiness disrupts rural life and displaces populations. In all these cases, enthusiasm for a new, promising tool mitigated against critical examination of the tool's limitations or harmful consequences.

The commodification of science produces a growing sophistication on the small scale and in the laboratory, but an increasing irrationality in the scientific enterprise as a whole. Economic concerns encourage the narrowness that runs through the major debacles of modern applied science. The quest for profits exaggerates the benefits of innovation, belittles the dangers, and claims more knowledge and control than is possible. Furthermore, as science becomes an increasingly commodified knowledge industry, the criteria of quality shift from validation by open and critical peer testing to those applied to any commodity. Does it sell? How profitable is it? Crucial knowledge is withheld in the name of proprietary information. The most egregious example is perhaps the tobacco industry's concealment of smoking's long-known toxic effects.

Any serious scientific method must correct not only individual error but also the prevailing collective biases. I would propose the following guidelines.

PREPARING FOR SURPRISES

Science must recognize Hegel's principle that the truth is the whole. A problem must be posed big enough to accommodate a solution. Even then, we must acknowledge that there will be surprises. The best that science can do is study the unknown by pretending provisionally that it is like the known. This is the case often enough to make science possible. But the unknown is also unlike the known, deeply enough to make science necessary. Especially in times of rapid change, there will be surprises. It is therefore useful to ask, how do other species cope with uncertainty? We must be prepared to be surprised, and we must develop science equipped to deal with the unknown, using the following approaches.

Detection of a Problem and Response

In order to be effective, the response must be rapid enough to prevent vast damage and to operate before things have changed again. Foraging ants return to the nest when the day gets too hot. Plants wilt when water loss becomes excessive. Insulin is released when blood sugar levels rise. Prey species flee their predators. In all these cases, the response comes after the changed condition has been detected. In order for this to be successful, detection must be accurate and rapid enough in relation to the duration of the threat. Otherwise organisms would always be responding to the previous situation.

In preparing for new disease problems, the public health system is concerned with efficient surveillance and reporting programmes. It is important that these programs operate rapidly enough compared to stay ahead of the spread of disease. Similarly, long delays in recognizing problems with toxic chemicals lead us to a dynamic state in which new pollutants are introduced into the environment as fast or faster than old ones are detected and removed.

Prediction

Many insects with short generations, such as aphids, prepare for winter by becoming dormant. However, the signal that starts dormancy is not winter itself, that is, low temperature or lack of food. Such signals would not be reliable, since cold spells could occur in mid-summer. Rather, the insects respond to the shortening days of autumn as a predictor of winter. But the silkworm responds differently. It has two generations per year. The first emerges in early spring when the short days indicate that it is the first generation and can have time for another generation. It is the information content of the environmental signal that matters. There is no necessary relation between the signal and the condition it predicts.[2] For human wellbeing we need both short-term prediction of the usual sort, for example: if a few cases of West Nile virus appear, there may be more. We also need longer-range predictions: if there are more mosquitoes, expect mosquito-borne disease. If we build dams, dig irrigation ditches, and kill mosquito predators, we may have more mosquitoes and mosquito borne disease. If we allow a depletion of biodiversity, exotic invaders are more likely to be successful. If we allow corporations to co-opt the environmental movement it will fail to protect humanity.

There is also need for long range, evolutionary prediction based on comparative epidemiology. We should identify those groups of insects that are potential disease vectors or major agricultural pests, that have the flexibility to extend their host ranges or modes of transmission or to adapt to new habitats. Most vectored human diseases are transmitted by flies, including mosquitoes, or by ticks. Among the plant viruses, the majority are transmitted by homopterans (aphids, scale insects, mealybugs). Like mosquitoes they are sucking insects that remove liquid from their hosts and return liquid to offset the vacuum their sucking creates. We should know which mammals are good reservoirs of viruses that could also infect people, which groups share more diseases with humans, and which kinds of diseases are likely to be shared. For instance, the human gut and the gut of a cow are quite different, but our lungs are similar, so we might be especially alert to the spillover of respiratory diseases.

Broad Tolerance

Plant breeders distinguish between "vertical resistance" and "horizontal resistance." Vertical resistance, provided by a single gene, gives complete protection but only to one pathogen type.

It lasts only until the pathogens overcome the resistance. Horizontal resistance usually depends on many genes. It protects against a broad range of enemies, and it persists because pathogens or insects have to do many things, more than they are capable of doing, to overcome it. In general, biodiversity provides horizontal resistance. It reduces the vulnerability of communities and is an important element of ecosystem health.

We need to design horizontally resistant systems that will defend against most diseases and poisons, even if we do not know what these might be or when they might appear. The vulnerability of our eco-social systems thus becomes a major target for research. It is especially important to take into account Schamlhausen's Law[3]: systems in extreme or unusual conditions or at the boundary of their tolerance are more sensitive to all environmental factors.

Poor and oppressed populations are more vulnerable to slight changes. This shows up in increased variability to health status. For instance, for any age group, blood pressure is more variable among African Americans than among whites, and life expectancy varies more in that group across cities. In evaluating the potential effects of some environmental or socio-economic change, we must look at the most vulnerable groups.

Prevention

Organisms can change their environments in ways that make potential threats less likely to arise. For instance, trees create the forest environment, which modulates the flow of water and slows down strong winds. Ants nesting in the soil create environments where temperature extremes are less frequent.

The precautionary principle encourages prevention as far upstream as we can reach. Instead of looking only at how to regulate an industrial toxin, we adopt a strategy that starts by asking, "Is this product necessary?" Then, if it is, in fact, necessary, we consider the best way to make and use it, with the least damage to the source of the materials, the workers who produce it, the consumers and the general environment.

Prevention includes doing the right kind of research so that practical and reasonable alternatives are available. The organization of research, the setting of priorities and the allocation of resources should be a matter of public debate. We must not continue to follow the spontaneous and haphazard course of investigations that lead to marketable commodities but ignore the complex relationships that might thwart the beneficial effects of invention, bring unexpected "side effects," and divert scientific attention from less intrusive, gentler technical possibilities.

Mixed Strategy

The best strategy combines all of these approaches. It prepares us to cope with the possibility that, even after we have used our best judgment and proposed the best solution we can find, we may still be wrong. It cautions us not to be caught up in technological fashions but to reserve the capacity to prepare for something different.

Rejecting Reductionism

The demand for a science of wholeness is a reaction against the very powerful but limited reductionism that sees the truth residing in the smallest pieces of reality. A science of wholeness does not reject reduction — taking something apart to see what it is made of — as a research tactic, but does reject reductionism — the illusion that once you have done that, the rest is an exercise for the reader.

The conflict between reductionism and a complex science of wholes has emerged from the domain of abstract philosophical contention, to become a practical political issue. It expresses the division between those who see the world as separate objects to be domesticated and processed for market and those who see the world as themselves and the places they live, between those who are reluctant to take responsibility for their actions beyond the most narrow immediate effect of their products, and those who see themselves responsible for their actions as they percolate through the world to the furthest detectable consequence.

Recognizing Connections

Things are more connected than we realize, even across disciplinary boundaries. Human biology is a socialized biology that varies according to our positions in the world. Real physiological differences exist among social classes in such things as the cortisol (stress hormone) cycle, immune system, and balance of excitatory and calming neural activity.[4] The common dichotomies of social/biological, genetic/environmental, physiological/psychological, and life style/social circumstance are false and misleading. The task of science in studying the complex eco-social phenomena of concern is not the assigning of relative weights to factors but to understand how the wholes fit together. These wholes include physical, biological and social processes.

Recognizing Bias and Partisanship

We are part of the system. The agendas and methods of science have their history, their strengths and their biases. Science has a dual nature as part of the generic progress of human understanding and the product of a knowledge industry that helps its owners maintain their prosperity, power and self-justification. Therefore, we need to look at our own biases as well as those of the fields we work in and not accept as given the rules that represent conventional wisdom.

It is important to acknowledge that science is often a battleground in which interests are defended in the guise of seeking truth. While striving for objectivity is necessary in science, neutrality is not. Just as industrial science serves industry, critical, dissenting scientists should acknowledge a frank and joyful partisanship: we do science in order to understand the rich complexity of the world; to have a deeper appreciation of the beauty and excitement of a marvellous, intricate and spontaneous biosphere and to cherish and protect it; and to benefit humanity, but especially the people most excluded from the benefits of our society.

Partisanship also implies a departure from the Earth Day injunction not to point fingers. The history of tobacco, asbestos, PCBs, pesticides, auto tires and many other industries show that corporations lie to protect profits. They mobilize staffs of public relations experts, lawyers and politicians to prevent interference with that quest for profit. While it is sometimes possible to demonstrate to an industry that improved environmental protection is also profitable, this is not always so and profit is still the goal of business.

So we are in a conflictive relationship with private, corporate industry. While ecology demands the limiting of growth to what is needed, corporations require growth. While ecology seeks to minimize inputs of energy and materials, industry seeks new products for turning resources into commodities. While ecology recognizes the unique qualitative properties of different species, habitats, resources and people, a commodified economy measures them all on the single scale of economic value and treats them as interchangeable. While corporations must seek to limit their responsibility to the narrowest possible effects of their actions, an ecological approach traces the consequences of industrial activity through the complex networks of the biosphere.

The corporate economy sees people as labour power or consumers, and spends millions to make us insecure enough to need its products, discover new uses for the raw materials corporations own, and invent new needs regardless of impact on health and well-being. Therefore, any suggestion of potential harm is regarded as "bad for the economy." But the ecological perspective sees resource use justified only insofar as it improves human life. Thus, we do have to point fingers and hone our science for battle.

DEMOCRATIZATION

Democratizing science would help to mobilize the collective intelligence of our species for the solution of shared problems. The discovery that every place is different, every forest and beach unique, every species special, means that there will never be enough scientists to analyze and study every phenomenon. We need everybody's imagination, experience and knowledge. Each time a previously excluded group — people of colour, women, working people — has been able to enter science, new insights have emerged that previous positions of privilege obscured.

Democratization has three elements: a democracy of recruitment into science, the popularization of science in a way that respects people's capacity as well as acknowledges their past limited access; and encouragement of nonprofessional participation in setting agendas, gathering and examining information, and analyzing results. Many examples illustrate the value of popular participation in science. Here are some:

- In environmental struggles such as those at the major corporate toxic spills at Love Canal in New York State or Woburn, Massachusetts, neighbours recognized a problem long before epidemiologists began to think about it.[5] The environmental justice movement identified patterns of discrimination and inequality. Groups such as the River Network monitor local pollution (www.rivernetwork.org).

• Ornithology has a long history of amateur participation. In just one Ornithological Laboratory survey of backyard birds, more than 16,000 people submitted more than 53,000 checklists reporting 4.5 million birds belonging to 442 species.[6] Despite uneven sampling and possible misidentifications, this mass effort gave a good sense of the status and changes in U.S. bird populations.

• In Cuba, epidemiologists organized teams to survey more than five million potential mosquito breeding sites around Havana, both to evaluate the effectiveness of control programs and to judge current problems. In 72 days the 2002 outbreak of *Aedes aegypti* was stopped by a combined efforts of some 11,000 people identifying and eliminating breeding sites.[7]

Biases in Experimentation

Currently accepted methods of experimentation and hypothesis testing create certain biases that are so standard procedure that they are seldom examined but rather are taken for granted.

A hypothesis is tested and shown to be likely only in comparison to certain other hypotheses. Therefore, the choice of which hypotheses to compare is crucial. When the chemical industry claims that new genetically engineered varieties are needed to save the world from hunger, they compare the expected yields of their products to doing nothing to improve yields. Or the industry may compare applying pesticides to letting the bugs run rampant. Such choices ignore gentle alternatives as diversifying plantings, encouraging predators and parasites of the pests, and strengthening the physiological resistance of crops through nutritional management of soil. Similarly, pharmaceuticals are tested against placebos or standard treatments but not usually against holistic approaches. The design of hypothesis testing requires a bold imagination to create real alternatives worth testing.

Healthy Populations

Statistical design is usually improved by reducing sources of extraneous variation. Thus laboratory animals are of a uniform genetic makeup, in good health and are raised under constant conditions. Human populations are selected with careful exclusion of people who may have extraneous conditions. They are matched for such things as age, sex, smoking habits, number of children, body weight, dietary habits, or anything else the investigators think might confound the results, provided the data are available. Often convenience, such as likelihood of being able to monitor people for a long time, determines the class or occupation of the participants. In occupational work, researchers want to examine people who have been at the same job for a long time. Thus, we miss the effects of a pollutant against a background of many pollutants, unstable employment, and varying economic conditions.

In order to avoid this type of error, studies must be conducted with the most vulnerable populations, often minority and poor communities. The variance of data exhibited in such studies is not simply noise interfering with detection of "main effects," but also an object of interest in its own right, an indication of stressful conditions leading to greater vulnerability. For instance, my colleagues and I found that when we grouped human settlements by size (central metropolitan area, smaller metropolitan area, smaller cities, and rural) across the populations within each category, the variance of mortality was greater for African Americans than for whites. African Americans are more vulnerable to economic differences and to differences in the structure of racism that has had much less effect on whites.[8]

Multiple Risks

In a population subject to multiple risks, statistical tests are less sensitive. For instance, suppose that in an exposed population a toxin causes 20 deaths per hundred and, in a controlled population, only 10. Then the relative risk is 2. But, if in addition to the toxin under study, there are 100 deaths due to other causes, the total deaths are 120 and 110. The relative risk has fallen to 1.1. Dioxin has not become less toxic to the exposed population; rather, the environment has become worse. Each environmental toxin helps to mask the effects of the others.

Sample Size

Statistical tests all depend on sample size. Studies often do not show sufficient numbers of positive results to achieve statistical significance. A lack of statistical proof is often misrepresented as evidence that there is no problem. Instead, the appropriate description of negative findings is that with this particular sample a suspected effect has not been detected. A more robust, interdisciplinary approach might detect a problem that a single statistical test might miss.

Confidence Levels

Estimates of the outcome take the form of a most likely effect, a comparison of difference (where zero means no difference, no effect) or a risk ratio, (where a value of 1 means equal risk with and without the suspected toxin). This is presented with a confidence interval. If the confidence interval includes zero (for differences) or 1 (for risk rations), we conclude that the suspected effect is not demonstrated. The press and industry then interprets this as proving there is no effect. But suppose that our risk analysis comes out 1.3±0.7 (with 1.3 being the relative risk and 0.7 the confidence level). That means the result is compatible with a risk level of 1 (no effect) or as much as a doubling of risk. It is even compatible with the conclusion that the substance is beneficial! However, in such cases of high uncertainty, a possible doubling of risk may be sufficient grounds for action.

Separate Variables

All statistical tests examine the relationship between some possible cause and an outcome of concern. They presume some model that relates the suspected cause to its effect. The model usually identifies a dependent variable or outcome, such as cancer, and independent variables that might influence it, such as pesticide use. The statistical analysis attempts to show whether or not there is an effect and how great it is. Then the model assumes that the prevalence of cancer is equal to the impact of a unit of variable 1 multiplied by the degree of exposure to that variable, plus the impact of a unit of variable 2 multiplied by exposure to this variable, and so on. If there is no impact, that variable contributes nothing to the prevalence of cancer.

Adversary Statistics

Adversary statistics is the use of statistics by opposing parties, each selecting new sets of data until they are satisfied with the results. They are fundamentally different from studies in which a design procedure is set in advance, followed, and conclusions drawn. If a contending party does not like the results a new study is done, and the process is continued until the results satisfy the sponsors. These are then reported as newer research that refutes "generally held beliefs."

Long after Gregor Mendel's classic experiments with pink and white peas that established the foundation of Mendelian genetics, it was observed that the data were suspiciously close to the 3:1 ratio that confirmed Mendel's hypothesis. Further research suggested that Mendel was so sure of his result that he continued counting peas until they came as close to the 3:1 as he could reasonably expect, and then stopped. This did not constitute cheating, since Mendel was not guided by statistical theory, but it was erroneous.

This revelation opened up interest in optional stopping and the field of "sequential analysis." The criteria for accepting hypotheses when we are free to carry out as many studies as we want are different from fixed-sample statistics in which it is decided before the study how much data to examine. This process is more complex than the now familiar optional stopping of sequential analysis, in which there is only one decision maker and no conflict of goals. There is at present no general statistical theory to address adversary statistics in which contending parties may carry out additional studies until they like the outcomes.

These considerations refer to duplications of previous research. A whole new domain opens up when continued research uncovers new harmful effects of chemicals, and the terms of the dispute evolve along with the information. As one question is settled, new ones arise.

Beside recognizing and addressing these standard biases, I would propose two further improvements in scientific method. First, incorporate nonprofessional data. A democratized science with non-professional data collection has special statistical properties. On the one hand, technical errors are likely to increase the standard error, expressed as the confidence interval within which the "true" number that we are trying to estimate will lie. On the other hand, the increased number of replications reduces the error. Thus if the error is increased ten-fold and the sample size k-fold, the new confidence interval will be multiplied by $\sqrt{10/k}$. And if k is greater than 10, the estimation is improved. We gain more than we lose by mobilizing community data collection. Moreover, if we want to determine trends, the heterogeneity of the data (times, local site conditions, and so forth) increases the variability of the data and reduces resolving power, but also increases the numbers of possible observations. Further, the heterogeneity itself is a virtue if the conditions under which samples are collected are recorded.

Second, historical analysis is a useful supplement to traditional testing. The history of most molecules in contention is that, over time, the number of known deleterious effects attributed to them has increased as we have looked more closely, tested for more effects, and understood more of the subtleties of interaction. Further, molecules in the same families often differ in the magnitude of their effects but tend to be qualitatively similar. For example, molecules in which the carbon atoms form a branching structure instead of a straight chain (phthalates, for example, or hydrocarbons combined with chlorine such as dioxins) do not occur naturally in animals. We have not evolved enzyme systems for dealing with them, but they are chemically active and can enter our metabolism. Therefore they should be suspect even without any direct evidence. Guilt by association is legitimate for molecules.

Understanding dynamic complexity is the central scientific problem of our time. In other words, we need to look at science itself as an object of study, a historically developed way of producing knowledge that creates a rich mix of insights and confusions. And our approach needs to be frankly partisan, rejecting the notion that feeling is the enemy of reason or that a commitment to human well-being is an enemy of objectivity. The suggestions outlined above would get us closer to a good, combative, perceptive scientific method that is more reflective of the complex, dynamic world in which we live and more supportive of precautionary decisions.

Three
Convergence of Development Goals

Human beings, individually and collectively, pursue many different goals at the same time. These goals may be independent of each other, come into conflict, or be mutually enhancing depending on the context, the available resources, the time frame for advancing toward these goals, conflicting interests, and their understanding of the circumstances. For instance, in the United States Afro-Americans and Latinos are among the poorest sections of the working class. Both need jobs. But because of capitalism there is unemployment, and the unequal distribution of wealth in society determines that there will be insufficient social expenditure to sustain the unemployed. Under these conditions, and because the politics of major cities is often organized along ethnic lines, the two groups may compete. But when the working class organizes in solidarity and there is united struggle for jobs and social support, the goals of the two are mutually supportive.

Or, as long as corporations threaten cities to relocate if environmental standards are enforced, the goals of employment and clean air conflict. But if cities were free to establish publicly owned industries or penalize companies that relocate to avoid environmental regulation, then the conflict is ameliorated. It is a working hypothesis for leftist politics in the United States that when two equally just goals are in conflict, we are asking too little and accepting constraints we need not accept.

Even in the absence of class conflict, scarcity creates conflicts among equally worthy goals. In present day Cuba, in the context of the blockade and the scarcity of urban transport, the need for urban transport conflicts with the need for clean air, and strict emission standards for buses in Havana are not yet possible.

A third kind of potential conflict arises because even when socially compatible goals are shared by the whole society, their implementation is assigned to separate agencies. They pursue their missions with single-minded devotion but without considering how their own work affects other societal goals. People trained as specialists may have difficulty looking at the whole.

It is the thesis of this chapter that under conditions of extreme scarcity and short-term urgency, the different social goals are pursued separately and may conflict. But as the urgency recedes, the time horizon broadens and the cultural level permits a broader vision. Then there is a convergence among revolutionary goals and their programmes. Their mutual synergy becomes manifest, the need for cooperation becomes more apparent, and the capacity to confront the complexity of the revolutionary enterprise deepens. The cross-linking among these programmes becomes both necessary and possible, and the responsible agencies must seek out cross linkages, integrating the revolution into a coherent whole. This convergence and integration is dictated by the processes themselves but can be retarded and distorted by pragmatist ideology or facilitated by Marxist thinking.

In what follows I will take agriculture as my point of departure and trace some of its cross links with other aspects of society, but especially with "environment" in the broadest sense.

CAPITALIST AGRICULTURE

It is generally assumed that agricultural activity aims at feeding people. This seems so obvious that it is rarely even acknowledged much less questioned. But if it were true, we would have to explain a number of anomalies. Why does hunger persist? Why is so much good fertile land not used to grow food? Why does food aid sometimes cause hunger? Why does India export wheat while village children are hungry? Why is international trade in agricultural produce dominated by non-foods (coffee, tea, sugar, cocaine)?

Is it due to a poverty of nature? In that case, why is it that during the Sahelian drought precious water was used to produce vegetables and flowers for the European market? There are other anomalies: if production were driven by nature, we would expect efforts to improve on nature, to protect the productive capacity of the land and to reduce the vulnerability to pests. But land is being lost to erosion, desertification, compaction and salinization.

Under capitalism, the relation between production and price is a negative feedback loop. If production were driven by fluctuating environmental conditions such as rainfall, low production would result in higher prices and higher production would give lower prices. We would expect a negative correlation between production and price. But in international trade the correlation is positive: high prices stimulate production and low prices discourage it. In the United States, grain yield depends more on the ratio of the price of grain to the cost of fertilizer than on rainfall. Agricultural production is driven not by human need or the productivity of nature but by the capitalist economy.

Or perhaps we are doing our best but do not know enough to produce more. But then why do some governments pay farmers not to produce? Why do the rich speculators buy up land around the large cities of Latin America, graze a few cows in order to call it a farm, and wait to resell to developers at much higher prices?

Perhaps the intricacies of relations among plants, soil, climate and microorganisms are just too complex for our science to understand? But then why is it that we have invented hundreds of pesticides and learned the intricacies of their molecular structures but do not even know what limits the populations of whiteflies or how to integrate livestock production with fruit growing? Why do we know more about soil additives than about nitrogen-fixing bacteria and the mycorrhizae that mobilize other minerals? And who we mean by "we"?

But the patterns of world agriculture, the array of inputs and technologies used in farming, the protection and degradation of crops, the unevenness of our knowledge about plants and animals, become intelligible if we assume an alternative hypothesis: most farming in the world today is the production of commodities, by the input of commodities and human labour. Although most of the world's farmers are peasants producing both for household consumption and the market, most of the world's farmland is dedicated to profit maximization through commodity production. Food is then a by-product of agriculture, not its purpose.

If we examine world agriculture as commodity production, then its major features are more easily understood:

All crops are interchangeable on the single scale of profitability. The most profitable crop will be preferred rather than any nutritionally optimal mix.

Investment in farming is measured against other possible investments such as land speculation, commercial construction, or banking.

Crops will be grown where they are more profitable than other crops or land uses.

Crops will be located where they are most profitable, resulting in regional specialization. The harvest will be sold to those who can afford to buy it, not those who need it most.

Profitability is uncertain as energy costs and markets shift from year to year. For slowly growing trees and for livestock this can thwart projections and result in cycles of boom and bust. Coffee is going through such a cycle now, with over-production aimed at a favourable market when the trees were planted, but production undermines current prices. Production for international trade is inherently unreliable.

There is no necessary relationship between the food value of a crop and its economic value.

There is no necessary relationship between the profitability of a crop and the environmental impact caused by its production. Crops, which are carried away almost completely, such as sugar cane, remove the nutrients from the soil and the field whereas a fruit tree harvest removes only a small fraction of the biomass each year.

Plants will be grown where most profitable, not where the plant is ecologically best adapted or the harvest most needed.

Human labour power is a cost of production. Therefore the laying off of workers whether or not there is other work for them is a saving, an increase in productivity.

Harm to the workers through accidents or toxicity from agricultural chemicals will be ignored and denied as long as possible, until the workers organize to demand safer working conditions.

Harm to the air, the water, and biodiversity will be ignored and denied as long as possible.

If the production system undermines its own capacity to continue producing, the producers can calculate whether it is more profitable to deplete the ecosystem as rapidly as possible and then invest the profit in something else, or preserve that capacity for long-term returns on the investment. Thus ranchers in the Costa Rican rainforest are content to clear the forest, raise cattle on a declining pasture, and abandon it after ten years with their investment richly compensated and the land in waste.

The inputs to agriculture — machinery, fertilizers, pesticides, and feed — are produced as commodities. In advanced capitalist countries they account for 80-90 per cent of the value added in food production, with only some 10 per cent added on the farm. Agricultural knowledge is produced when the chemical and engineering corporations can turn it into marketable commodities. This leaves us with a pattern of ignorance and knowledge that is not dictated by nature but by the commercialization of research.

Advocates of the ecology movement often see this pattern of agriculture as irrational. And from the perspective of humanity as a whole it surely is. But each kind of society has its own rationality which makes perfect sense to its practitioners. For instance, a major criterion in economic calculation is "efficiency." Efficiency seems like a simple, objective idea that can be applied to all situations. But what is the efficiency of agriculture? In the Bible it is considered a blessing when the sowings are "returned manifold." In land-hungry Europe, efficiency might be seen as yield/hectare. In the United States, with abundant land (in the past) and sparse labour, efficiency was calculated as yield per labour day. According to this measure, labour saving mechanization which dismisses workers is highly "efficient," especially when they organize. We still hear boasting about how one American farmer feeds 40 people. But it is now one farmer, four or five production workers off the farm making the inputs the farmers consume, and an equal number handling and processing the yield after harvest. On the other hand in Cuba, where urban agriculture employs some 10 people per hectare, this would seem grossly inefficient. However, in the wake of the Special Period the creation of jobs and the provision of food were both urgent social goals. Some 300,000 Cubans, many of them women, are engaged in urban agriculture, each producing vegetables for 36 consumers. For the present this is very efficient socially, but as more capital becomes available to create productive employment in other sectors it will no longer be the case and technologies that save labour may become more desirable. Finally, ecologists have begun to calculate efficiency as energy harvested per energy invested.

When we examine the efficiency of agricultural production we have to include changes in the capacity to produce. A cement factory that over-fulfilled its production targets but had to grind up the walls and floor to do this would not be regarded as very efficient. In agriculture too we have to include changes in soil organic matter, the turn-over rate of leaf litter, the number of species of predatory coccinelids and other indicators of ecosystem well being.

Each of these measures of efficiency makes sense in its context but none are universal. When we begin to consider the social "efficiency" of agriculture, its "yield" in the satisfaction of human need, we have to take into account its impact on a wide range of economic, social and ecological criteria including food value produced, economic viability, improvement of soil and biodiversity, employment, conservation of water quality in the drainage basin, retention of cooperative members in the countryside, educational advancement, and general social and political participation of the farm community. But that is a task for socialism.

PRAGMATISM, MARXISM AND THE CRITIQUE OF AGRICULTURE

So far in the world, socialist development has been attempted only in poor countries. Under conditions of extreme scarcity and short-term urgency, different social goals confront each other externally, as alternative claims on resources in short supply. Land in parks is land not producing food. Teaching is not picking fruit. Standing trees are not timbers for houses. To some extent equally important goals even compete in the sense that leaders charged with meeting a specific goal tend to regard other, equally important tasks, as obstacles to their mission. Administrators are often impatient with theory and dismiss long-term goals as idealistic. Not only are material goods in short supply but also educated people who can cope with more complex problems, and the scientific knowledge that makes a whole-system approach both necessary and possible may not yet be available. Preparing for the future can detract from the present.

It will be proposed in the section on convergence and disintegration that as the extremes of poverty are overcome, it becomes possible and also necessary for the different socialist programmes to converge, overlap, support each other mutually. This may happen at first spontaneously, because it is the obvious thing to do, but will also be reflected intellectually as a more self-conscious integration of Marxist thought with planning and invention.[1]

Under conditions of extreme scarcity and short-term urgency, pragmatism is a congenial ideology. Anything that meets the most urgent needs of the people should be done. Pragmatists do not look for any indirect consequences of their actions, any harm they might cause. Philosophical questions are regarded as distractions, criticism as obstruction. Socialist pragmatists borrow much from capitalist pragmatism. They share common beliefs and methods and a pride in being "practical, realistic" people. They also regard science and technology as neutral instruments that harm or benefit according to how they are used, rather than as products of specific social relations developed or neglected according to the needs of their owners.

But despite superficial similarities, capitalist and revolutionary pragmatism are fundamentally different. Capitalist pragmatism is driven by the demand for profit. Therefore the harm that a technology might cause to workers at the point of production, to the population as a whole, or to long term sustainability are not only ignored but hidden, denied, actively defended by batteries of hired experts, and eventually incorporated into their budgets as business expenses. The tobacco industry in the United States is a classic example of suppressing information, research whose conclusions are paid for in advance, false claims, intimidation of critics and thwarting of regulation. Therefore conflicts about the health impacts of technologies, although they take the form of scientific debates, are really episodes of class struggle.

Revolutionary pragmatists also defend their approaches vigorously, but in this case the debates are real debates between differences of opinion, of philosophy, and are rooted to some extent in the division of labour, but not a fundamental clash of interests. Therefore scientific evidence carries real weight and can eventually prevail.

We dwell on the conflict between pragmatism and Marxism because it helps to understand the zigs and zags of socialist development, in particular around agriculture and the environment. The pragmatist argument is developmentalist. It sees development as "progress," as "modernization." Modernization is seen to advance along a single axis from less developed to more developed, and the task of the less developed is to catch up with the more developed as quickly as possible. Pragmatists see technology as the driving force, and seek out the technologies that seem to "work" without examining the process of development of technology and the real choices that are available. Planners in poor countries are easily dazzled by the abundance produced in the rich capitalist countries and have an inordinate admiration for capitalist ways of technology and management. Socialists have even used the term "business-like" to mean practical, realistic, no-nonsense, unsentimental and efficient, rather than the "business" we really know: greedy, intolerant, short-sighted, secretive, callous, and dishonest. They reproach ecological criticism as "idealist," a sentimental attachment to some golden age that never really existed, while ecologists respond that it is the height of idealism to imagine that we can pass a resolution and have nature obey. Some pragmatists also criticized the emphasis given to public health in Cuba as too expensive, a luxury for so poor a country. They see themselves as the more consistent materialists. They see industry as the model for agriculture. They expect a modern agriculture to emerge from backward peasant and plantation agriculture through the following transitions:

- From labour intensive to capital intensive production through chemicalization and mechanization.
- From small-scale *minifundia* to large-scale giant extensions of land so as to take advantage of the economies of scale especially with heavy machinery and irrigation.
- From the patchwork of heterogeneous land use of peasants to the homogeneity of the specialized industrial farm, more suitable to heavy specialized machinery and chemicals and easier for a simplified top-down management that can prescribe uniform practices over the whole field.
- From traditional practices (often dismissed as superstition)

to practices guided by modern scientific knowledge, especially by scientific knowledge of the very small (molecules).
• From vulnerability in the face of the uncertainties of nature to control of nature.

Since the 1970s there has been a growing criticism of this pathway of development coming from different sources with different concerns. Consumers concerned with food quality and safety, farm workers exposed to pesticides, environmental groups wanting to protect potable water, indigenous communities seeing their whole productive base destroyed, the organic farming movements in a number of countries, farmers wanting a healthful life style, conservationists concerned with biodiversity, religious and secular advocates of the simple life, leftist critics of capitalist development have all converged in the recognition that the prevailing pathway of industrial, high-tech agriculture is self-limiting and destructive.

Capital-intensive agriculture is very expensive, making a country very vulnerable to losses in foreign exchange. It develops plant and livestock varieties that produce a lot but depend on inputs of energy, chemicals, rigorous weeding, antibiotics, imported feed, and for poultry controlled temperature. Unequal access to these inputs promotes a growing inequality in the countryside. When these become unavailable the system collapses.

It undermines the productive capacity of the land, leading to increasing erosion, compaction and salinization of the soil and depletes or contaminates the aquifers.

Large extensions of a single crop make the system more vulnerable to the uncertainties of the weather and to highly mobile pests and pathogens, and have fluctuating needs for labour and water. They therefore provide variable income to the producers. They ignore the unevenness of landscape, the variations of microclimate that could be used for producing diverse crops. It also often creates the habitats for the vectors and pathogens of human disease. And despite frequent claims, large-scale monocultures are not more productive than the alternative: a mosaic landscape of mixed land use and polycultures within fields. Ivette Perfecto has found that a coffee plantation cultivated in the traditional way under shade also preserved some 90 per cent of the invertebrate diversity of the natural vegetation of the region.

Regional specialization increases the risks that the whole year's production could be destroyed by a hurricane with a diameter of 200-300 km. and in normal times depends on efficient, often refrigerated, transport.

Industrial monocultures are biotic deserts, poor in soil invertebrates and microorganisms that help preserve fertility, poor in natural enemies of pests, and unobstructed to those pest species that can handle the extreme environment.

Industrial agriculture also harms the broader environment. Run-off of fertilizer gets into the ground waters. Nitrites are carcinogenic for humans; fertilizer feeds blooms of algae that deplete the oxygen in lakes and rivers, killing fish and insect larvae. Some of these feed on mosquito larvae, so that modern agriculture may increase diseases such as malaria and rift valley fever. The soil particles washed into the lakes and streams increase turbidity, harming fish culture. Pesticide runoff kills wildlife and accumulates also in the human body. On the broadest geographic scale it contributes to global climate change.

Ecological agriculture rejects the dichotomy between peasant and scientific knowledge. The call to "learn from the people" is an important antidote to the arrogance and condescension of progressivist science that often come from the dominant class or who seek their validation in the university centers of Euro-North America rather than from their own people.

Listening to the peasants is especially important with ecological technologies that are site-specific. During the Cultural Revolution in China, the validation of peasant knowledge helped to mobilize the collective intelligence that had been so often passive. But it can also lead to serious errors, mostly in the direction of pragmatism, short time horizons and dismissal of theory. Peasant knowledge, like all knowledge, has to be studied scientifically.

Marxist epistemology teaches us that all knowledge comes from experience and reflection on that experience in the light of previous knowledge, belief and interest. Therefore the world-view of any social group is some combination of rich insights and incredible blindness. The collaboration between people of different histories requires an understanding of the strengths and typical errors of all participants.

In general peasants are acute observers of their own environment insofar as it is relevant to their practice. They distinguish among plant species that have to be treated differently, no matter how similar their appearance. And they lump species that can be treated in the same way. Usually children, guided more by curiosity, recognize more different species than do adults, and independent farmers know more natural history than do farmer labourers.

But not everything is visible. Peasants have very subtle knowledge of processes and things that exist on the time and spatial scales of everyday experience that they can see—plants, insects, and soil texture. But they do not experience or know about photosynthesis of growth substances. They understand, or once understood, water management. In ancient Mesopotamia water management was very sophisticated but nobody understood soil chemistry. In the high Andes, Aymara farmers raised potatoes on raised beds above canals, at an elevation too high for commercial potato growing. The canals serve as sinks for heat during the day, protecting the crops from night frost.

But peasant knowledge is local, not grounded in theory, and not easily generalizable. The Santiago de Cuba meteorologist Fernando Boytel once told me of a valley in Oriente where the peasants assured him that trees grow into the wind. Everything we know about plant physiology tells us that trees should grow more on the side away from the drying effects of the wind. But in fact the trees in that valley grew toward the wind! It turned out that because of the topography and geography of the region, sunlight and wind came from the same direction, and the effects of light on photosynthesis overwhelmed the drying effect of the wind. The observation was accurate, the interpretation in error.

Once we know what the typical patterns of insight and error will be both for the peasants and for ourselves, we can all be aware of them openly. Then agricultural scientists and peasants can meet as equals in a process Paolo Freire called communication as against extension. It is important that the pattern of error be understood. Although the same error is unlikely to be repeated, there are kinds of errors. In medical practice there are now studies as to the reasons for deaths during surgery or for misdiagnosis of a disease or delayed recognition of an epidemic. In works of applied science and national or local policy, it is not enough to acknowledge that errors have been made. We have to understand what the major kinds of errors are, why they were made, and what delayed their correction.

The boast that agricultural technology gives us control over nature and protects us from its vagaries is naïve. What it does is displace uncertainty: imported feed protects livestock from the variability of rain-fed pasture. But as we learned during the Special Period in Cuba, it makes production more vulnerable to the uncertainties of prices and world politics. Nitrogen fertilizer protects wheat growth from uneven soil composition, but makes the Argentine wheat plant sensitive to the price of West Asian oil.

Sometimes, opponents of planning argue that because of the universality of uncertainty the hopes for a planned economy are illusory. There are two classes of uncertainty: those that arise from external events such as climate or international politics and others that are generated within the system. For instance, modern industrial plants have many safety devices and redundancies to respond to failures of the parts. The probability of any one failure is small, but the number of parts is so great that we have to calculate: vanishingly small probability times a very large number of opportunities becomes nearly inevitable. And when modern systems do fail, the very connectedness of the system allows the failure to spread. The impacts are far more catastrophic than the more frequent but small failures of older systems. This is true of nuclear power plants, electrical grids, and genetically modified organisms. The other kind of breakdown is a result of the internal dynamics of a system when we try to control it too closely and we respond too strongly but too slowly to correct failings so that they are amplified instead.

But another response is possible: if even the most modern technologies are vulnerable to the uncertainties of the world, perhaps we have to study uncertainty itself as a fact of life. One way of doing this is to ask, how do other living things, with two billion years of evolutionary experience behind them, cope with a changing and often surprising environment? We have found four modes of coping, not mutually exclusive:

Rapid Detection and Response to a Change

This depends on the accuracy of detection, the rapidity of response, and the effectiveness of the remedy. Detection and response form a negative feedback loop in the system. Negative feedback can be stabilizing, but if the delay is too long and the response too strong but brief it leads to oscillations in which we are always doing the right thing for the previous situation. For instance, in the community of HIV/AIDS sufferers in California, everybody knew somebody dying from AIDS. Therefore educational efforts were successful and AIDS declined. But then people no longer knew AIDS sufferers. The urgency to take precautions diminished, and the next cohort of vulnerable people experienced an upsurge of AIDS. The same thing may happen with dengue control: an outbreak of dengue leads to active mobilization and effective removal of breeding sites for mosquitoes. The very success then reduces diligence, other priorities demand attention, and the mosquito population can recover. *Conciencia* itself diminishes over time (like isotopes it has a half-life). Thus detection and response requires persistence in sanitary measures and the education that makes this possible.

A similar problem arises in pest management: the outbreak of a pest may panic the producer into using pesticides. These may successfully reduce the pest, but in so doing increase other pests later by killing their natural enemies.

Prediction

Here an indicator is used for a problem that has not yet arisen. We know now that dry weather increases the likelihood of a whitefly outbreak or that moist winds can spread the blue mold of tobacco. Insects in the temperate zone use the length of the daylight to respond to winter before it gets cold. Birds use the shortening day length to prepare for the fall migration. There is no necessary relation between the indicator to which the organism responds and the condition it is preparing for. We need systems for short, medium and long-range prediction for weather, pests and diseases. For instance, observed cases of malaria predict the future cases. And abundance of mosquitoes predicts malaria. Heavy rains predict mosquitoes. Clearing of forests and use of pesticides that kill the enemies of mosquitoes predict mosquito breeding.

Climate change confronts agriculture with changes that can harm the present day crops. For instance increased wind velocity damages bananas, high humidity reduces the yield of potatoes, droughts increase white fly populations. Awareness that any major change in land use is a potential change of epidemiology allows longer-range prediction while evolutionary studies allow us to anticipate the adaptation of new pests or disease of our crops or of people. Thus a multi-layered prediction system is needed.

Prevention

Here an organism or population reaches out into the environment in such a way that the problem does not arise. People stop building homes on vulnerable beaches; children are immunized even if there is no outbreak. Education reduces exposure to HIV/AIDS. Mosquito breeding sites are removed.

Broad tolerance

Or what plant breeders call horizontal resistance. Then it is not necessary either to detect or predict since the system will resist whatever comes along. For instance when there is a whole community of natural enemies of a pest, we do not have to monitor all their populations but only assure that we do not destroy them with pesticides and that there are sources of nectar for wasps, straw for spiders, shade for lion ants, nesting sites for bird and bats. Resistance is a property of the whole ecosystem, not any one beneficial predator alone.

Mixed Strategy

And finally there is the mixed strategy. Since none of the defences are sure to work, we are best off with a combination, multiple lines of defence. In science and politics it encourages us to guarantee that while a broad consensus can decide on the main direction of our efforts we always have to leave room for other experiments, just in case we turn out to be wrong or because we know that new situations can make our present successful measures less effective. For instance, immunization is perhaps the major public health strategy against infectious disease today. But we have to imagine that the age of immunization, like that of antibiotics or pesticides, is a successional stage in our relationship with the microbial world and may outlive its capacity to protect us. If we take seriously that everything changes we have to look beyond even our most successful present approaches for what we will do next, when they are no longer as effective as they are now.

Pragmatists now acknowledge at least some of the harm caused by the single-minded pursuit of production. But they argue, isn't this the price of progress? Don't we have to accept these costs in order to feed the billions who are now hungry? This is the argument of the agrochemical corporations who have never cared about the world's hungry but also of the pragmatists who do, who are overwhelmed by the visible suffering of poverty. Isn't the protection of nature a luxury for a poor country? "As long as children are starving, I cannot care about the ivory billed woodpecker (*carpintero real*)."

Marxists have been active participants in the critique of industrial agriculture in two ways. First, we recognize the dual nature of science: it is both part of a generic unfolding of human understanding of the world and the product of particular societies whose rulers need some parts of knowledge and ignore or suppress others. The character of modern science is determined by the needs of the owners of the knowledge industries, the social groups recruited to manage or conduct science, and the beliefs and traditions of the scientific communities since the seventeenth century. We know that these communities have their own biases and procedures for excludinq unacceptable ideas. We are aware that surprise is inevitable in science, that theories, like isotopes, have half-lives, and that the fashionable ideas of the moment are often those that promote domination. It provides us with a skeptical view of the ideas that dominate world science that allows us to make good use of the achievements of that science without national chauvinism but without being blinded by them.

The second is more specific to the problems of science and development. The major failings of modern scientific approaches have come about because problems were posed too narrowly, too statically, and undialectically. Public Health was caught by surprise by the reemergence of infectious disease as a major world health problem even after the proclamation of the "epidemiological transition" to chronic disease. It failed to take into account that each change in ecology, vegetation, demography, social conditions, and social relations is also potentially a change in epidemiology. It ignored the evolutionary potential of pathogens and their vectors to invade new habitats and hosts. It did not pay attention to the rise and spread of new diseases among animals and plants, part of the same universal adaptive mode of parasitism. It assumed that economic development would lead to affluence for all and therefore to the diseases of the rich. It was unprepared for the unexpected, that pesticides could increase pests, that hospitals can be foci of infection, that food aid could increase hunger. Marxists have challenged the prevailing reductionist and static biases of modern science, calling for a dynamic holistic approach to our complex world, but without falling into the mysticism that is often attached to that critique. Therefore we have been in the forefront of the development of agro-ecology.

It is the argument of agro-ecology that the self-serving claims of the corporations are not true, that the high-tech industrial approach that swamps nature expecting to produce more is counterproductive, that a gentle, thought-intensive and complex production system will be highly productive, more stable in adversity, more humane, sustainable, and compatible with more of the general social goals of socialist development.

AGRO-ECOLOGY

Modern agro-ecology was led by the quest for alternatives to pesticides, especially to DDT. In the agricultural universities of the United States the pest control courses were dominated by the industries that offered research money and support for students. But students rebelled. During the legal hearings about DDT in the 1970s, faculty of the University of Wisconsin testified in favour of the use of this pesticide while students spoke up against it. At the University of California at Davis and at Cornell students demanded courses in alternative agriculture. Ecologists thwarted in their own country found more willing collaboration in revolutionary Cuba and Nicaragua. Although the use of natural enemies was a long familiar technique, its adoption as the major strategy was new.

The theoretical basis for this work is clearly predator/prey relations. The predator/prey relation is a negative feedback: any increase in the pest provides more prey for the predator; the predator increases and consumes more prey, restoring the population. But if we apply pesticides to this system, the pesticide kills both kinds of insect. The predator is harmed in two ways: it is poisoned directly and its food supply is also poisoned. The prey suffers from direct poisoning but benefits because its predator is killed. Therefore it may even increase! There are further complications introduced into this model when there are more than one kind of predator, some of which also use other prey, when there are delays in the response of the predators, when climate affects the dynamics of both species or when the predator can finds the prey on some host plants but not others. A modern pest management scheme does not rely on a single natural enemy but on managing the whole ecological community with multiple lines of defence.

The Breeding of Resistant Varieties

This is usually interpreted to mean that an herbivore cannot feed on the plant or that an infection is not successful. However when we consider that our objective is to protect the harvest, not the plant, it is possible to broaden the definition of resistance. For instance, if a bean plant develops for a few weeks before it is infected with bean golden mosaic virus (BGMV) the infection does little harm. Therefore resistance might consist of a shorter critical period for infection. If we are able to delay the arrival of whiteflies with the BGMV, then even a minimal shortening of the critical period is "resistance." Other, non-infected whiteflies can feed on an infected plant after the virus has had a chance to reproduce, and then that whitefly is infective. Therefore any genotype that lengthens the latent period of the infection may prevent secondary infection prior to harvest. This is another kind of resistance. There is also regional resistance. In the whitefly-BGMV system, some crops support the whitefly but are not infected. Others, such as the common *frijol*, are susceptible but are not a good reproductive host for the whitefly, but permit it to feed and move on. Thus a field belongs to one of 8 classes according to whether it does or does not support feeding, reproduction of whiteflies and infection by virus. (Of course, if the whitefly cannot feed it cannot reproduce or transmit virus, but an evolutionary alternation might allow it to feed where it previously could not, and then whether it can reproduce or transmit becomes relevant). From the point of view of controlling this disease, a region is a mosaic of different kinds of fields, and a region as a whole may be resistant to BGMV if the virus cannot be transmitted continuously from crop to crop in that region.

Agricultural genetics focuses on breeding varieties for yield, tolerance of the environmental conditions, and resistance to pests and diseases. There is not yet a programme for breeding predators and parasitoids of the pests. We have to consider the breeding of support species as a new direction for genetics.

The physical arrangement of plantings of the crop with companion crops that can confuse the herbivores and reduce feeding such as beans with tomatoes, and the use of repellant or trap crop. Polyculture research was initiated at Guira de Melena by INSAV. There is now a large literature of useful crop combinations.

Generalized predators that are present even when the pest is not there, and can prey upon it as soon as it arrives but its population does not respond well to increases in the pest. These include ants, spiders, lacewings, dermopterans and beetles among invertebrates but also birds and bats. Our major intervention, perhaps after an initial introduction, is not to destroy their habitats.

Specialized parasitoids of pests, mostly wasps and flies. They appear only after the pest is present but then can increase in response to the pest population. In order to have them when they are needed we can raise them in production centers (CREE) for release or else provide hosts for them on alternative crops or protected areas within the farm or region.

Diseases of the pests, usually fungi such as Beauvaria or virus diseases of insects. They are most effective when the best density is high enough for them to spread. Some of them can be natural parts of the ecosystem while others still have to be raised separately and released.

If the pests escape all of these defences, we have the choice of direct intervention through the application of non-toxic products such as neem or we could accept that the insects share our environment and will occasionally exact a tribute.

A second pathway of ecological agriculture is the improvement of the soil. The goal here is a stable complex ecosystem rich in organic matter, nitrogen fixers, and mycorrhizae, with the communities of fungi and invertebrates that recycle litter and suppress outbursts of pathogens. The use of earthworms, bacteria and fungi can improve the fertility of the soil while the selection of crops for different root geometries can reduce competition and increase the mobilization of minerals from deeper layers. A soil rich inorganic matter can hold more moisture, is less vulnerable to fluctuating rainfall, and demands less irrigation.

All of this is still on the scale of a single field. But ecological development of agriculture also has a larger spatial dimension. The farms of the future will not be specialized enterprises with a single product but a mosaic of different land uses arranged in patterns most suitable to the topography and the needs of different crops for labour and inputs, and the direct interactions of adjacent plots. For instance, bananas provide sufficient shade to maintain permanent populations of the *hormiga leona, Pheidole megacephala,* that is a very aggressive predator against pests of sweet potato. Therefore if the bananas are planted next to the sweet potatoes in alternating strips no wider than 40 meters the ants can forage among the sweet potatoes and need not be introduced. The permaculture movement is leading the way in this direction. Each patch in the mosaic can contribute both through its own harvest and through its service to the rest of the farm.

Forests yield lumber, charcoal, fruit, nuts, coffee and honey. They are also the refuge for birds, bats and other natural enemies of pests and for pollinators of fruit. They modulate the flow of moisture so that there is less variation in available moisture, and through friction they reduce wind velocity that offers some protection to the crops. Forests create a special environment around their edges to a distance of about ten times their height that are rich in wildlife and suitable for some crops. They offer shade for farm workers and are a recreational resource for the country. These forests would vary in size from several rows of trees to many hectares. Cultivated forests can maintain a significant fraction of the biodiversity of the natural forests of the region and contribute to the belts of protection for the fully protected areas.

Pastures raise livestock, provide manure, retard erosion if well managed, maintain a balance between grasses and legumes that can enrich the soil, and provide nectar for the parasitoids of crop pests.

Similar analysis can be applied to crops of different sorts and to farm ponds. In addition to their agricultural uses, forests, wetlands and ponds can support resident or migrating wildlife and thus contribute to national biodiversity.

A farm with a mosaic of land uses answers the conflict between the needs for large scale planning and ecologically scaled patches: the unit of planning, the economic unit over which income is shared and distributed, and the unit of production do not coincide. The unit of planning has to be large enough to be in balance with the needs for water, machinery and labour while the unit of production is more determined by the mobility of pests, the topography of microclimates and the ranges of interactions of plants. The unit of redistribution includes the whole mosaic so that peasants are not penalized for growing low profit, high utility crops.

At present we have the knowledge to make a good start in designing a mosaic structure of mixed enterprises. The limiting factor is likely to be the problems of management of so many different activities at once. This can be accomplished by a nested array of decision making, with each problem being solved at the most local level compatible with the problem and subject to the requirements from the more inclusive system. For instance national policy may require that a region produce both for export and local consumption. At a more local level the different land uses can be combined according to local requirements. Suppose for example that we have two crops and a limited amount of water. We estimate the water requirements of each crop for each month and then ask, what proportion of the land devoted to each crop will minimize the deviation between required water and available water? In order to do this we have to combine the mathematical tools for optimizing with constraints, with the specific knowledge of the people who are working the land. Of course in real practice we would have many constraints and many degrees of freedom (resource use and practices). Decision making must lie as close as possible to the people doing the work. The broad autonomy of the producer has implications not only for the effective administration of agriculture but also for the health, social integration, and demographic stability of the rural population, as will be discussed below. A prerequisite for its development is the educational and political level of the rural population.

Finally we have to examine the large geographic scale of agriculture. In a countryside organized mainly into mixed enterprises, there will certainly be some regional specialization. But it is not true that regions should specialize according to comparative advantage as recommended by bourgeois economics. This would increase vulnerability to disasters, both natural and economic. Rather, each region should have a mix of production for export and for its own consumption, and a variety of basic crops, belonging to different plant families and ecological categories, to guarantee an adequate diet. The mix should include short and long cycle crops, water-requiring and drought resistant, grasses, legumes and composites, deep- and shallow rooted.

In the main, agricultural land has to be part of a pattern that includes national protected areas, local buffers along rivers or coastlines, and population centres, balancing demand on resources, availability of labour, and protection of landscape and climate. And since the point of farming is to feed people, the levels of production of meat, milk, grains and vegetables has to be linked with an assessment of the evolution of the Cuban diet as guided by our understanding of nutrition. Such an outcome can only come about through the integration of all major national projects. As such a project evolves, the landscape of the whole country will acquire a unique pattern so that a flying saucer from Alpha Centauri, swooping over the earth, will be able to look down on Cuba and say, "There is socialism down there on that island!"

CONVERGENCE AND DISINTEGRATION

A feature of socialist development in Cuba is that as we get beyond the immediate short-term urgencies, different separate programmes designed to meet different revolutionary needs begin to converge into a coherent whole, their conflicts diminish and the possibilities of mutual enhancement emerge. Their potential conflicts diminish because there are more resources available, the longer time horizon that allows the feedbacks and mutual enhancement to operate, and the education and awareness that permit the connections to be sought after and understood. Short- and long-term perspectives require dialectical thinking and the coordination of many state and non-governmental organizations. For instance the commitment to health for all is centered in MINSAP (Ministry of Public Health) but increasingly requires actions in *organismos* responsible for agriculture, environmental protection, scientific research, education, urban planning and other spheres. It will soon have to confront the restructuring of work as a response to an aging population and the role of stress in the genesis of disease so that input from the unions will become necessary. Programmes for conservation and protection of biodiversity have to link up with agriculture. Urban planning, whose first goal is providing housing for the population, has to take into account also urban agriculture, the physical environment, the climate of cities, the social integrity of neighbourhoods, and in the end the whole urban-rural metabolism.

Of course, in all societies the different aspects of the whole interact. But before the appearance of socialism the interactions were spontaneous, each part having its own relatively autonomous dynamics unconnected to larger social goals. The outcome was its own unique pattern of relations with the rest of nature, land use, disease spectrum, in which the impact on human welfare was a by-product, a side-effect.

This convergence and integration in socialism arises both directly out of the problems and opportunities of development and indirectly through the conscious application of Marxist insights. It is a consequence of the contradiction between the demand for a rising standard of living and the requirement of sustainability. This is resolved by taking the rising standard of living to mean primarily a rising quality of life rather than the unlimited growth of consumption of energy and matter, with an emphasis on collective consumption. As material conditions progress beyond immediate survival, the temporal and spatial horizons expand toward the long run, where all aspects of society interact and can reinforce each other. At the same time, an increasingly self-conscious Marxism makes a strategy of integration, wholeness and connection seem obvious and provides the tools to confront that complexity.

On the other hand, when a revolutionary society reverts toward capitalism it fragments into its parts. Whole sectors of the Soviet economy faced the rest of society increasingly as capitalist corporations even before the massive plunder and privatization of the 1990s, interacting with each other only externally, through the market. Soviet ecology also declined. It had been in the vanguard of world science in the 1920s. It gave us the concept of the biosphere, posed the possibility of the origin of life from the chemical evolution of the primeval earth, lead in soil science and biogeography. But as capitalist relations reappeared even in state enterprises, ecology starved through a lack of application to the development strategy and the withdrawal of resources, and was reduced to isolated actions such as organizing to protect Lake Baikal. It wilted intellectually. Environmental protection survived in part through fines imposed for polluting, but the large enterprises simply budgeted the payment of fines into their 5-year plans as production expenses. Although in 1932 Soviet scholars startled the International Congress of History of Science by showing the social determinants of science, by the 1950s science and technology were treated merely as a factor of production. In the last years of the Brezhnev leadership, the USSR relied on the idea of a politically-neutral "scientific and technological revolution" to save a stagnating economy. Faced with real difficulties and an anemic Marxism, Soviet leaders borrowed from the intellectual tools of the encircling world to design policies based on pragmatism, geopolitics, and liberal consumerism. The restorers of capitalism faced a disarmed and feeble resistance.

The opposite experiences of the USSR and Cuba show something about the role of ideas in development. A formal commitment to Marxism does not by itself determine development although it may promote or retard that development. The interaction goes in both directions. The decline of socialism in Europe also undermined Marxism, reducing it from a vibrant radical critique of all that is, to a rigid dogmatic apologetics or futile rearguard actions honoured on state occasions. In Cuba, Marxism did not create the integrated pathway of development that is now emerging but participated in it, and is in turn nourished by the emerging socialist relations. In particular, it was an active presence in the conflict between pragmatist "modernizing" and dialectical views of development.

THE RESTRUCTURING OF AGRICULTURAL LABOUR

People spend perhaps a half or a third of their waking time at work. The environment at work therefore has great importance for their health and quality of life although it is not usually regarded as a question of "environment." The transformation of work will be arising in Cuba, perhaps first in industry. But the transformation of agricultural work will not be far behind, and offers an opportunity to explore our notion of the convergence of different social necessities as a characteristic of socialist development. It enters the collective agenda in four contexts: agricultural productivity, health, demography and social/ political participation.

The new ecologically-guided agriculture is highly site specific. The configurations of the landscape, local and even micro-climate, geology and soil types, past history of land use, the natural flora and fauna of a region, all combine to create a unique situation that has to be managed in a special way and emergencies dealt with by means that are suitable for that place. This cannot be done by spraying thousands of agronomists over all of Cuba. The people who are engaged in production have to be able to evaluate the situation, know who to consult, make the decisions as to what course to follow, and then carry them out. This requires a combination of physical and mental labour and the autonomy to make decisions.

Research in industrial health shows that the prevalence of chronic conditions such as heart disease is very much related to the social environment at work. Robert Karasek and his collaborators have shown that stress on the job, coming from the challenges of work, is actually good for health if the worker has the autonomy to confront those challenges. But if high stress is combined with low autonomy, it creates anxiety and heart attacks. Low levels of challenge and autonomy are depressive. Further, people whose work is challenging and who enjoy a high degree of autonomy also engage in more activities outside of work. They participate more in social, cultural, political and recreational activities. And this comes about relatively rapidly, within a few months or a year of changing jobs. Thus flexible, diverse work which is challenging and with broad room for decision-making is important for the health of workers and for maximizing collective participation in the running of society. One final note here: what constitutes the right levels of stressors and autonomy will not be the same for all people. As the economy grows with full employment and relatively equal wages, people will actively chose their employment partly for job satisfaction that will include the comfortable levels of challenge and autonomy.

The Cuban *campesino* is now too educated to aspire to wielding a machete for his or her whole life. As job satisfaction emerges as a major determinant of where people live and work, the drift to the cities will become stronger unless farm work becomes more supportive of the full development of the person.

The population is aging. The combination of low birth and death rates guarantees that the average age will increase further and a smaller fraction of the population will be in the optimal age group for strenuous physical labour. One solution under capitalism is the importing of workers from poorer countries who can be paid less. This has given rise to the pattern we see in Europe and North America of immigrant labour being both welcomed and persecuted in a context of racism. Another solution is to increase the productivity of labour, which usually means the intensification of exploitation and the increased investment in energy and materials. Some improvements in productivity do not require such measures but rather more thought-intensive use of labour and skill. However sooner or later the question will arise of raising the age of retirement.

With work as most people experience it, such a measure would be highly unjust. For the majority of workers, retirement is liberation. It is the loss of pensions, income and health care that forces so many workers in the United States to continue working after age 65 at jobs that undermine their health even more. But for professionals and intellectuals retirement is often felt as punishment. In the United States it was a victory when we won the right to continue working in the universities as long as we are able to. The difference between the two cases is that professional work enjoys a high degree of autonomy and free movement among the diverse tasks of experimentation, teaching, and learning.

Thus the transformation of work in agriculture serves the needs for ecological production, improved health, retention of people in the countryside, changing demography, and social participation. It is this convergence of distinct societal goals that makes the transformation of agricultural labour necessary. The rising educational levels of workers and of planners and the awareness of the role of convergence in developing socialism are making it possible.

THE ECO-SOCIAL DISTRESS SYNDROME

The Eco-social Distress Syndrome is a multilevel pervasive dysfunctional relation between our species and the rest of nature. It includes the using up of non-renewable resources, over-exploitation of renewable resources, undermining of our major life support systems (agriculture, forestry, fisheries and pasture). It pollutes our environment with a historically new chemosphere of some 100,000 synthetic molecules most of whose effects are not yet studied. It threatens the biotic diversity of the biosphere. It changes our relations with the microbial world, giving rise to new infectious diseases, alters our demography and leaves us with a pattern of knowledge and ignorance that is inadequate to confront the complexities of our new situation.

It is not the first crisis that we have faced in the last 50,000 years. Other civilizations have faced social and environmental crises and have responded in one of three ways: they have moved someplace else and continued their old way of life; they have changed their way of life and continued in the same place; or they have failed to do either and we know of them through broken pottery, arrowheads, and grave sites. Sometimes they have compensated for a declining production by aggression and plunder. Nothing is more dangerous than a declining economy with a first rate military.

But the present crisis is different in being global, reaching deeper into the earth and higher into the atmosphere, affecting more aspects of our lives, its effects are more long lasting and less reversible. It is both a general crisis of our species and, because it is arising after 4 or 5 centuries of capitalism, it is also a specific crisis of the capitalist system.

One aspect of this crisis is the conflicting demands for a rising standard of living, equity, and sustainability. If present consumption patterns of the rich countries continues, and if developing countries aspire to catch up with Euro-North America in the use of energy, matter and area, then the earth cannot support us for very long. But if these three just goals conflict, we are asking too little. In this case, survival and equity become possible, but only if we understand the rising standard of living to be mostly in terms of the quality of life rather than consumerism.

The criticism of consumerism cannot be based on the asceticism of the affluent. We have to recognize that at present billions of people want more goods. But we can analyze why.

First there is the real need of people in poverty to have more and better food, housing, clothing, transportation, and health care. This can be accomplished when the GDP (or actually the better measures that are being explored now) reaches the equivalent of some $5,000-$10,000 per capita.

Second, there are the necessities created by a capitalist economy that must expand to survive. Automobile transportation, public or individual, really has become a necessity for many people because of the way urban development determined by real estate values separates the locations of residence and employment and creates high-density settlement, while polluting industries make distance from the workplace essential for health. At present public telephones are disappearing in the United States in order to make cell phones another "necessity." The competition among capitalist corporations gives a priority to speed. Much of the air travel in the world today and much of the hotel accommodations are a consequence of "business"—meetings, sales efforts, last minute shipping, while much of the construction is not for housing but for offices.

Third, with accumulated capital creating excess productive

capacity, sales effort commands a growing proportion of corporate expenditures. There is constant pressure to buy and to seek the personal validation that is denied people in an individualistic competitive society by accumulating possessions. Advertising aims at making people feel the need for more and more different kinds of products.

Finally, it seems to be the case that the consumption patterns of the rich and powerful acquire a prestige far beyond any intrinsic value of the objects. When Babylon conquered Judea and carried off the ruling class to Mesopotamia, the Judeans were dazzles by the splendors of the "world capital" and many of them turned down the opportunity to return home 70 years later under Cyrus the Great. In Jesus' time, Rome was the place to be. Herod spent years in Rome learning Roman habits, going to Roman parties, and meeting all the right people. He then tried to bring the delights of Roman civilization to ancient Israel. Now it is the USA. Everything from Coca Cola to baseball caps and CD's acquire a spurious virtue because they are from the USA. Even scholars seek validation by publication in American journals.

Thus a change in consumption patterns requires a change in patterns of land use and settlement, as well as a culture of collectivity and mutual caring.

Once that becomes our goal, we confront both a political and a scientific agenda. The political agenda is being conducted worldwide and is well known. The scientific agenda leads us to ask, what patterns of landscape assure sufficient food, protect the water and air and soil, provide a life-enhancing habitat for us in supportive communities, preserve species diversity, design livable cities? It is here that the convergence of the various socialist goals becomes imperative.

Four
The Science of Dharma and the Dharma of Science

The struggles around the nature and value of science can be understood as a three-cornered struggle in which the contenders are (1) a pre-capitalist holism which is hierarchical, static or ahistoric, reactionary and mystical; (2) a capitalist-era liberal, rationalist, secular, reductionist, instrumental scientism; and (3) a post-capitalist, dynamic, anti-hierarchical materialist dialectical holism.

This tri-partite classification is, of course, an abstraction from a more complex reality. Individuals form their own belief system in the context of their society and location in that society but also from more idiosyncratic experiences. Schools of thought borrow and are influenced by each other. Post-colonial societies often combine modern globalized capitalist relations with pre-capitalist modes of power and join their willing subordination to the global corporate system with a petulant cultural nationalism.

Opponents of the dehumanizing instrumental reductionism often hark back to romanticism. I will deal with some of these complications later. Like all abstractions my own are ultimately false, but before "ultimately" comes around they may be useful. The case for an abstraction is not that it is a flawless description of the world but that it grasps some central relationships and thereby advances our understanding, and that it leads to liberating practice.

In her recent work, philosopher of science Meera Nanda focuses on the struggle between the pre-capitalist mystification and modern science.[1] She is especially concerned with the adoption by the right-wing Hindutva movement of a mix of modern instrumental science with a mystified anti-science based on a forced interpretation of the Vedas. She is certainly aware of the limitations of modern science, its increased commodification, its corruption for profit or career advancement, its development of those areas of knowledge that serve the rich and powerful. But she sees these as defects that can be remedied by the ordinary mechanisms of peer review and open debate. On the other hand, she sees mystified nationalistic science as an essential ingredient of Hindu reaction and therefore a major obstacle to the development of a humane, modern, secular, technically advanced and liberal democratic India. "Hinduism-as-science is a part of the cultural project of modernizing without allowing the rationalism and secularism of the Enlightenment to challenge the traditional cultural values. . . . The danger of Hindu Science is that it will further entrench the holistic, organismic worldview as our national ethos . . ." (pp. 22-23).

Nanda's defence of modernist science does not make the naïve assumption of a free-floating objectivity: "This is as good a place as any to clarify that those like Souroush, Popper, and many post-positivist philosophers of science who believe that objective knowledge is possible are not saying that scientific knowledge is free from cultural assumptions, gender biases, metaphysics and such. All they are saying is that it is possible — albeit not always easy — to identify and put these assumptions, biases, etc. to a systematic test the reason and experiment. The danger of *social constructivism and the postmodern theories of science is that they deny that we can ever break out of the prison of our myths, bias, and cultural assumptions* . . . " (p. 135; emphases in original).

Nanda uses "holism" in a special sense. For her, it is the fusion of the material with spiritual worlds and the natural world with the social under the domination of the spiritual, mystical, mental domains. For example,

> • Hinduism, for Nanda, "is a cosmopolis in which two of the most inegalitarian ideas of nature — namely, nature as a hierarchical chain of being and nature as an organism — have combined to produce a highly inegalitarian social philosophy" (p. 154).
>
> • "*All* pre-modern cultures see the society as an organic whole in which natural events have a moral import, and conversely moral behaviour has an effect on the course of nature. The much vaunted 'holism' of non-Western traditions and has nothing non-Western about them: they're simply non-modern ways of knowing which abounded in the West until they were discredited by the combined assault of capitalism, the Reformation, the Scientific Revolution and the Enlightenment. Holism is simply another name for not separating out what actually belongs to the cosmos and what belongs to our polis and its traditional myths" (p. 155; emphasis in original).
>
> • "It [dharma] makes hierarchy an obligation imposed by the order of nature" (p. 156).

The villains of the piece are reactionary Hindutva nationalists, neo-Gandhians, postmodernists, eco-feminists, social constructivists, and a generic left whom she lumps as reactionary modernists. According to Nanda, they reduce modern science to an ethno-science of the West, while they look at the idealized Indian community as a dharmic unity.

Against this holism, Nanda sees dualism as a virtue. A dualist epistemology allows the worlds of science to develop free of the demands of religion and myth, in a way analogous to Prussian dualism which allowed a domain of relatively free scientific inquiry in exchange for staying away from the domain of politics. But the holism that is her target is quite different from the materialist holism which is a criticism of reductionist scientism. Materialist holism is an essential ingredient in the struggle against reactionary and mystical holism.

In political terms, Nanda sees the agenda for those invested in Indian politics as a completion of the bourgeois revolution. The desired outcome of this process is a modern India with a technically sophisticated ruling class integrated into the intellectual world of global capitalism: "before the liberal enlightenment regime of economic and civil liberties can be radicalized — as the Marxists desire — they must first be allowed to win legitimacy over and against the dharmic common sense of Indian society" (p. 118).

Nanda faults the critics of modern science on several grounds. First, they object to the separation of culture from nature, the knower from the known, matter from spirit, reason from myths and emotions. Second, they claim that rationality is an imperialist imposition, that women and non-Western people appreciate inter-connections and think in wholes. Third, Nanda rejects the 'staunch anti-Americanism' of the holists, their epistemic charity which asserts that all claims to truth are equally socially constructed and none can claim to bring us closer to truth.[2] Fourth, she rejects the call to cherish and learn from the people and the defence of even self-destructive customs against what are called elitist and rationalist criticism.

Revolutionary Marxists can only admire and support Meera Nanda's courageous resistance to the reactionary mystifications of Hindutva nationalists, the barbarities of caste and sati, and post-modernist nonsense. But there is also another front: the struggle against capitalist science. Other chapters in this book detail aspects of the post-capitalist critique of scientism. Here we only review them briefly.

Marxist holism starts from Hegel's dictum that the truth is the whole. It is a warning that if problems are posed too narrowly they end up attributing explanation to external events. We further claim that many of the major failings of contemporary science have come about because problems have been posed too narrowly. Two examples will suffice:

> (1) The doctrine of the epidemiological transition ignored history, ecology, and evolution. It looked at human disease separately from the diseases of plants and animals. It believed that the way to confront disease was through the search for magic bullets, drugs that could be marketed as commodities. And it ignored the epidemiological impact of inequality and poverty. Public health was caught by surprise with the emergence of new infectious diseases and the resurgence of old ones that we had thought already conquered.
>
> (2) The Green Revolution aimed pesticides at pests without considering the natural enemies and competitors of these pests embedded in communities of organisms. It ignored the social consequences of expensive high-tech inputs and the destructive consequences of heavy machinery, large doses of chemical fertilizer, and grand extensions of monoculture. Programmes for economic development failed to take into account their ecological and social impacts.

A common theme to these failings is a systematic reductionism that posed problems too narrowly, bounded by unstated constraints, and as static. Hegel's dictum is a warning that there is more out there. We must always ask in studying a particular problem "where is the rest of the world?" It is a further warning that what we have not taken into account can, and eventually will, overturn our theories.

Marxist holism does emphasize connection among phenomena. Deforestation alters the populations of mosquitoes and therefore the prevalence of malaria. HIV/AIDS in Southern Africa has an impact dependent on land tenure. New varieties of wheat may undermine the economic independence of women. The resurgence of rabies in Israel is linked to an increase in foxes caused in part by the destruction of the Palestinian economy, which encouraged people to raise chickens for food. The chickens are also fox food, and the removal of jackals by the Israeli authorities eliminated the major competitor of foxes just as the Intifada increased their food supply. Foxes are the major reservoir of rabies in Israel. Our emphasis on connections among phenomena does not mean that we can cast spells to get justice but that we must trace the connections even among phenomena that are not obviously related.

Dialectical holism does not reject reduction as a research tactic but reductionism as a philosophy of nature. It is certainly useful and often necessary to know what something is made of. But is it is an illusion to expect that knowledge is sufficient to understand the whole. It is a physiological fact that a pesticide can kill an insect in a bottle. But that is not equivalent to the ecological claim that the application of the pesticide in the field will improve the well-being of farmers.

As historical materialists we do claim that knowledge is a social product. But we do not stop with the naïve claim that received knowledge is false or that some knowledge is wiser than other because it comes from nicer people. Rather the claim of social embeddedness is a demand to study how that knowledge is produced. We claim that all knowledge comes from experience and reflection on that experience in the light of previous knowledge. Therefore we have to examine the domain of experience that gives rise to knowledge. Learning from the people does not mean that the less educated person wiser, or that "what the ancients say" is a proof of validity. We ask which ancients, why was what they said recorded and why millions of other ancients left no trace in the records.

Each social location gives rise to its own unique combination of insight and blindness and it is the task of the historian of science to discover what they are. Nonprofessionals often have a subtle highly specific knowledge of their own surroundings, but only of those objects on the size scale of daily experience, limited to locality which depends on the mobility of people and ideas, and limited in time by the social memory. Scientists approach the same problem with comparative information and general theory that has often abstracted away the complexities of the particular. Respect for national cultures and traditions serves as a counterweight to the arrogance of imperialist globalization. But national and traditional beliefs also come from stratified societies with clashing interests that have a stake in the content and accessibility of knowledge. The result is that ancient, like modern, views of the world are a real mixture of accumulated knowledge and what might be legitimately labelled as superstition. Ancient peoples looked at the sky and labelled the stars and constellations in very different ways according to their culture. But beliefs are not arbitrary. In the end nature intrudes. All peoples could recognize a year of some 360 days and navigate by the differently described stars. Those plants, which have wide geographic distribution, are also incorporated into traditional healing systems in similar ways in very different cultures. The objective world is viewed by different peoples according to their own cultures that, to prevent total caprice, impinge on the conclusions. And traditional agriculture combines the real experiences of polyculture, rotation, encouragement of the natural enemies of pests and other elements of modern ecological agriculture with fantastic ritual means for protesting against the weather and pests they cannot control.

"Epistemic charity" is not a blank cheque to the past but the acknowledgement that sometimes, old knowledge is profound where modern instrumentalism sacrifices the long run to immediate gain, and that the wholesale dismissal of traditional knowledge remains a part of colonialist arrogance.

Without giving any particular ancients an epistemic privilege we recognize that each social group has its own perspectives which colour the way in which it sees the world. For instance primatology as a science was transformed by the entry of women who immediately saw through the sexist understanding of troops of baboons or prides of lions.

We even link nature and morality, not in the magical way of the Hindu Right but through the working hypothesis that all theories are false which promote, justify or tolerate injustice.

We endorse Nanda's support for ancient progressive traditions instead of those that came out on top in the ancient struggles and helped to implant a durable and vicious exploitative system. From our point of view the social construction of science is a tool for focusing scientific methodology on ourselves in order to look critically at the agendas of our own sciences and our own prevailing theories. It also lets us challenge the shared biases of the sciences. It allows us to question the gatekeepers of science who allow only some to participate in research, to publish and to teach.

Nanda is familiar with the fallibility of science but expresses confidence in its self-correcting capacity: "If there was a dogma of the Enlightenment, it was that there would be no dogmas, no a priori truths and no privileged Sources of Affirmation. All dogmas could be queried by private citizens, who will have the right to come together in the public sphere, as equals, to pursue truth through open critical debate" (p. 112). These ideas are, of course, marred in practice "but an exclusive hypercritical concern with these contradictions can blind us to the momentous implications of winning public legitimacy for new forms of public reason" (ibid.)

But it is not that these ideas are "marred in practice." Rather the whole evolution of the production of science as a knowledge industry and of knowledge as a commodity undermines these ideas "in practice." The cost of research makes the independent testing of results impractical. Commercial and military secrecy and political blockades block the sharing of data. Economic interest can determine if research is published or suppressed. Scientists may be pressured by their employers to adapt their results to the employer's interests. And the scientific community has biases and traditions for dismissing uncomfortable ideas as "not mainstream," "unproven," "quackish," "ideological" or by other terms from a long list of designations designed to exclude the unacceptable ideas. The bigotries of the science tribes are as dogmatic as any tribal mythology. The classical methods of error correction serve well for avoiding idiosyncratic mistakes. Laboratory glassware should be clean, experiments should have controls, the double-blind experiment protects us from the unconscious biases of the experimenter, and statistical techniques protect us from attaching too much significance to random events. But the shared biases of the scientific community are defended as furiously as a pre-capitalist community might protect its gods. It is these shared biases that turn scientific rationalism into scientism.

I am sure that Nanda is aware of the oppressive aspects of present day high-tech corporate science. She nevertheless chooses to dismiss the criticism of bourgeois science as a secondary problem compared to the struggle against the reactionary modernism of the Hindutva fundamentalists. She is more concerned with the urgency of resisting the drift toward the right in Indian politics, and separates indigenous reaction from its imperialist allies. In order to do this, she lumps Marxists with all sorts of post-modern anti-rationalists who smooth the way for the Hindutva reactionaries.

Is she right? Is the threat of a fascistic regime that combines pre-capitalist mysticism and anti-rationalism with modernist technology so overwhelming, oppressive and imminent as to justify allying with the scientistic rationalists to stop that danger? And is the defence of modern science with all its scientism the best way to resist reaction?

It is not. The golden age of "free" science (for some) is gone, strangled by monopoly capital's increasing ownership of the knowledge industry, the reduction of science to commodities and scientists to scientific labour power. The scientistic errors that lead to the failures of public health to control preventable disease, for agriculture to seriously confront hunger, for development to produce prosperity and for industry to nurture a habitable earth warn us that these are not secondary faults. The defence of reason cannot win if based on the flawed scientism that prevails among the modernizers. When Christian fundamentalists attack "humanistic science" as also socially determined, we cannot fight back by saying "No—it is your nonsense which is socially determined while ours is pure and objective!" Rather, we agree with them and then go on to show what social determinations are at work. When they say that science is not infallible, we agree that in the long run all theories are false, but in the immediate some are relatively true and others are terribly flawed from the start. The battle has to be fought out on a terrain enriched by the philosophy of science. The science we defend must be critical, dialectical, and rooted in a historical perspective.

Then who are our allies, who our adversaries? And how does Meera Nanda fit in? A foreigner looking from a distance cannot know the political landscape of India intimately enough to identify the various contending parties or the degrees of urgency of different struggles. However some observations are possible.

First, for all its nationalism the Hindu fundamentalists are not anti-imperialist except at the level of cultural rhetoric. The globalizing transnational corporations and their governments are not allies for rationality. The turn to the right in India, for all its specific features, is part of a general drift to the right that afflicts the world's capitalist democracies. Democratic forms are being hollowed out in favour of aggressive nationalistic regimes with power shifting to the corporations directly, their collective agencies, and an increasingly domineering executive. In this terrain, at least in Latin America, the survivals of indigenous democracy with its emphasis on participation and the building of consensus rather than the adversarial politics of capitalism is a rich starting point for the socialist democracy that is beginning to take shape. The "pro-Americanism" that seems a liberal antidote to national chauvinism is really an ally of imperialism and assists the penetration of the hegemonic power. Resistance to the technocrats is resistance to both enemies.

Meera Nanda is our ally in the struggle against the Hindu right, the traditional oppressive sexism and inequality, the mystification of nature. But she is not our ally, apparently, against the corporate modernizers or against scientism. Vandana Shiva, on the other hand, is an ally against the technocratic globalizers, against scientism, but not, apparently, against rural mysticism. It is the nature of coalition politics that allies can sometimes be adversaries, adversaries sometimes allies.

In entering coalitions, we have to distinguish between what to insist on as a principled prerequisite for alliance, and what to advocate without requiring the same of our partners. We should not demand of allies more than they are prepared to agree to.

There is much mysticism in the environmental movement; we advocate a more materialist understanding but do not demand it of our allies.

Most of our adversaries today should be our allies in the future. Some of them have been and will be again many times over. It is always important to treat them honestly and with respect and be ready to learn from them. Just as Marxism was nourished by German philosophy, English political economy, and French socialism, so now our Marxism is and should be enriched from the feminist, environmental, and ethnic liberation movements.

Five
Is Capitalism a Disease?

The scientific tradition of the "West," of Europe and North America, has had its greatest success when it has dealt with what we have come to think of as the central questions of scientific inquiry: "What is this made of?" and "How does this work?" Over the centuries, we have developed more and more sophisticated ways of answering these questions. We can cut things open, slice them thin, stain them, and answer what they are made of. We have made great achievements in these relatively simple areas, but have had dramatic failures in attempts to deal with more complex systems. We see this especially when we ask questions about health. When we look at the changing patterns of health over the last century or so, we have both cause for celebration and for dismay. Human life expectancy has increased by perhaps thirty years since the beginning of the twentieth century and the incidence of some of the classical deadly diseases has declined and almost disappeared. Smallpox presumably has been eradicated; leprosy is very rare; and polio has nearly vanished from most regions of the world. Scientific technologies have advanced to the point where we can give very sophisticated diagnoses, distinguishing between kinds of germs that are very similar to each other.

But the growing gap between rich and poor make many technical advances irrelevant to most of the world's people.

Public health authorities were caught by surprise by the emergence of new diseases and the reappearance of diseases believed to be eradicated. In the 1970s, it was common to hear that infectious disease as an area of research was dying. In principle, infection had been licked; the health problems of the future would be degenerative diseases, problems of aging and chronic diseases. We now know this was a monumental error. The public health establishment was caught short by the return of malaria, cholera, tuberculosis, dengue, and other classical diseases. But it was also surprised by the appearance of apparently new infectious diseases: the most threatening of which is AIDS, but also Legionnaire's disease, Ebola virus, toxic shock syndrome, multiple drug resistant tuberculosi, and many others. Not only was infectious disease not on the way out, but also old diseases have come back with increased virulence and totally new ones have emerged.

How did this happen; why was public health caught by surprise? Why did the health professions assume that infectious disease would disappear and why were they so wrong? In fact, infectious disease had been declining dramatically in Europe and North America for the last 150 years. One of the simplest kinds of predictions is that things will continue the way they have been going. Health professionals argued that infectious disease would disappear because we were inventing all kinds of new technologies for coping with them. We now can carry out diagnoses so rapidly that some diseases that might kill a person in two days can be identified in the laboratory soon enough to permit treatment. Instead of spending weeks culturing bacteria, we can use DNA to distinguish between pathogens that may have very similar symptoms. More importantly, we had developed a new arsenal of antimicrobial weapons, drugs, and vaccines, as well as pesticides to get rid of mosquitoes and ticks that are disease carriers. We came to understand that, through mutation and natural selection, microorganisms could present a recurring threat. We assumed that whatever the microbial changes, the disease-causing mechanism would remain the same while we developed ever-newer weapons against it. It was, we believed, a war between us and the microbes, in which we would have the upper hand because our weapons were growing stronger and ever more effective. Another cause for optimism — at least this was the argument put forth by the World Bank and the International Monetary Fund — was that economic development would eliminate poverty and produce affluence, making all the new technologies universally available. Finally, the demographers noted that while most infectious diseases are most deadly to children, we have an aging population, so the proportion of missing from this hypothesis was that one-reason children are so vulnerable is that they have not developed the immunities that will contract diseases at an older age. Indeed, some diseases, like mumps, are more serious in adults than they are in children.

So what was wrong with our epidemiological assumptions? We need to recognize that the historical mindset in medicine and related sciences was dangerously — and ideologically — limited. Nearly all who engaged in public health prediction took too narrow a view, both geograpically and temporally. Typically, they looked only at a century or two instead of the whole sweep of human history. Had they looked at a wider time-frame, they would have recognized that diseases come and go when there are major changes in social relations, population, the kinds of food we eat, and land use. When we change our relations with nature, we also change epidemiology and the opportunities for infection.

THE PLAGUE IN EUROPE

Plague erupted in Europe for the first time in the sixth century during the decline of the Roman Empire under Justinian. Europe suffered from social disruption and declining production. The sanitary facilities of the great ancient cities were crumbling; under those circumstances, when plague was introduced it swept through the population with devastating effects. Plague reappeared in the fourteenth century during a developing crisis of feudalism, causing a population decline even before plague became widespread. The standard history of this plague occurrence is that sailors landing in ports along the Black Sea brought plague with them from Asia in 1338; it then spread westward and, in a short time, reached Rome, Paris, and London. In other words, plague spread because it had been introduced from elsewhere. But it seems more likely that plague had entered Europe many times before but really didn't take off. It only became successful when the population became more vulnerable, when the human ecosystem could not confront a disease spread by rats at a time when the social infrastructure that would have controlled rats had crumbled.

AN ECOLOGICAL PROPOSAL

When we at look at other diseases, we see that they rose and fell with historical change and circumstance. So, instead of a doctrine of the epidemiological transition, which held that infectious disease would simply disappear as countries developed, we need to substitute an ecological proposal: that with any major change in the way of life of a population (such as population density, patterns of residence, means of production), there will also be a change in our relations with pathogens, their reservoirs, and with the vectors of diseases.

The new hemorrhagic fevers appearing in South America, Africa, and elsewhere almost all seem to be related to increased contact with rodents that humans don't normally meet, caused by the clearing of land for the production of grain in particular. Grain is also rodent food; rodents survive by eating seeds and grasses. When a forest is cleared and grain is planted, we also eliminate the coyotes, jaguars, snakes, and owls that eat rodents. The net result is an increase in rodent food and a reduction in rodent mortality. The rodent population grows. Now these disease-deliverers are social animals. They nest and build communities; when a new generation emerges, the young adults go out looking for homes elsewhere — often wandering into warehouses and people's homes, facilitating the transmission of diseases.

Another human activity, irrigation, is especially related to the breeding of snails, which transmit liver fluke disease, and mosquitoes, which spread malaria, dengue, and yellow fever. When irrigation proliferates, as it did, for example, after the construction of Egypt's Aswan Dam, habitats for mosquitoes were created. Rift Valley fever, which had occasionally erupted in Egypt, can now be found fulltime. The development of giant cities in the third world has created new environments for the spread of dengue, transmitted by the same mosquito that transmits yellow fever (Aedes aegypti). It has adapted to life around the edges of cities. A poor competitor against other varieties of mosquito in the forest, these mosquitoes are able to breed in abandoned lots, in puddles, water barrels, and old tires — in the special environment that we create in the giant cities in the tropics. Dengue and yellow fever are particularly threatening because of the growth of urbanization in the tropics with megacities like Bangkok, Rio de Janeiro, Mexico City, New Delhi, and others with populations of ten to twenty million. As human population grows, there are new opportunities for diseases. For instance, you need a few hundred thousand in a population before it can sustain measles. If there are fewer, measles can infect the entire population; those who survive will be resistant. But if there aren't enough new babies to maintain the disease, it will disappear and have to be reintroduced. But in a population of a quarter million people, there will be enough new babies who are not resistant that the disease can sustain itself in the population. Consider this: if we know there are diseases that require a quarter million people to be self-sustaining, what diseases will emerge in crowded populations of ten or twenty million? Clearly, as life conditions change so do opportunities for disease.

Yet another kind of myopic thinking in the public health community arose from the fact that doctors are concerned with human diseases but have not paid much attention to diseases of wildlife or of domestic animals or plants. Had they done this, they would have had to confront the reality that all organisms carry diseases. Diseases come from the invasion of an organism by a parasite. When an infection takes place, it may or may not produce symptoms. But all organisms deal with parasites and, from the point of view of the parasite, invading an organism is a way of escaping from competition in the water or in the soil. For instance, the bacteria that causes Legionnaire's Disease lives in water. It is found all over the world but is never very common because it is a poor competitor. It has very finicky dietary requirements, so normally humans don't encounter it. However, it has two things going for it. It can tolerate high temperatures and is resistant to chlorine. It withstands chlorine by hiding out inside an amoeba. In a convention centre, hotel, or truck stop, water is both heated and chlorinated. And if it's a good hotel, we may find a showerhead that gives a fine spray of tiny droplets, perfect for carrying the bacteria into the furthest corners of your lungs. What we've done is to create the ideal environment for Legionnaire's Disease. The chlorine and the high temperature kill their competitors, the remains of which form a coating on the inside of the pipes that is marvellously rich in the food that the Legionnaire's bacterium loves.

If we look at other organisms, we see a constant jockeying for position between parasites and hosts. The more common a species is, the more attractive it is to new invasions by parasites. Humans are very common, and thus offer wonderful opportunities for invasion. When we observe disease patterns, we see that cholera, for example, spread from the Eastern Hemisphere into the Americas, entering Peru and then travelling up to Central America. But a similar route was followed by a disease of orange trees, by viruses of beans and tomatoes, as well as by wildlife diseases. What we see, then, is a constant coevolution between pathogens and hosts across all animal and plant life rather than a situation unique to humans. Surely, we would have a much better understanding of potential dangers if we understood human illness from this perspective.

TRANSMISSION OF DISEASES

What kinds of insects spread viruses to people? Nearly all of them are mosquitoes or flies, or belong to a second group which includes ticks, fleas, and lice. These are the two main groups that overwhelmingly spread human virus diseases, even though there are hundreds of thousands of other kinds of insects. There are very few diseases spread by beetles, none that I know of by butterflies or dragonflies. Why? Are there circumstances under which they might become transmitters of diseases? Among plants, the major distributors of plant viruses belong to a totally different group of insects — to aphids. However, both groups have similar mouths and subsist by sucking liquid from their hosts: the mosquito sucking blood, the aphid sucking sap. If you have ever sucked something through a straw, you know that after a while a vacuum builds up and in order to be able to continue slurping the liquid, you have to be able to return liquid. Similarly, the salivary glands of mosquitoes and of aphids return liquid to their hosts when they take up the blood or the sap, and in that liquid you find the viruses. That's why when we study viruses, we look at the salivary glands of mosquitoes, or of ticks, or anything else. We can begin to encounter these generalizations when we step back from looking at the particular details of a given disease and try to get a broad picture. But this wasn't done.

THE FAILURE TO STUDY EVOLUTION AND SOCIETY

Another kind of scientific narrowness — really a self-imposed intellectual constriction — is the failure to study evolution. Evolution tells us immediately that organisms respond to the challenges of their environment. If the challenge is, for example, an antibiotic, organisms will respond by adapting to those antibiotics. In agriculture, we know of hundreds of cases of insects that have become pesticide-resistant; in medicine, increasing numbers of micro-organisms have become resistant to antibiotics meant to fight them. Some microbes have become resistant to antibiotics even before they are used! This happens when an antibiotic is released on the market with a new trade name, but in fact is hardly different from its predecessor. It may look different, but if it acts on the bacteria in the same way, it will be met by the same defences. It is not enough to look at the agent of disease; we have to look at what makes populations vulnerable. Conventional public health failed to look at world history, to look at other species, to look at evolution and ecology, and, finally, to look at social science. There is a growing body of literature that says that the poor and oppressed are more vulnerable to nearly all health hazards. But we still don't recognize class differences in the United States. Researchers discuss differences in income or a mother's education level or even socioeconomic status. But U.S. epidemiology does not deal with class, even when class is the best predictor of life expectancy, of old-age disability, or the frequency of heart attacks. As a predictor of coronary disease, it is better to measure class position than to measure cholesterol.

OTHER EXPLANATIONS

Why do we wear these intellectual blinders that have so hobbled the study and practice of U.S. public health? First, there is a multiplicity of long-term intellectual biases. Take, for example, American pragmatism. Americans pride themselves on their practicality. "Theory" is almost a dirty word. When we are overwhelmed with the urgency of a population that is sick, of kids that are dying, it becomes a luxury to ask about evolution. This overwhelming sense of urgency is one of the reasons why doctors don't look at diseases of tomatoes, don't ask about competition between different kinds of mosquitoes, and certainly don't look at historical factors. There is an inevitable tunnel vision built into the urgency of carrying out applied clinical or epidemiological work.

A second reason is the western scientific tradition of reductionism, which says that the way to understand a problem is to reduce it to its smallest elements and change things one at a time. This is very successful when the question is, "what is this made of?" Then we can isolate it, cut it out of an organism, put it in the blender or under the microscope. In fact, we have been marvellously successful at identifying what things are made of. That is why we have had a growing, if irrational, sophistication about small phenomena and events throughout the whole of scientific enterprise. Why is it we are so successful at giving individual emergency treatment and so ineffectual in stopping or preventing malaria, in anticipating its return, or dealing broadly with the health of whole populations? We are marvellously successful at breeding a wheat plant that can better use nitrogen to produce more grain but much less successful at alleviating hunger in the countryside.

FOUR HYPOTHESES

So the typical failure has been a refusal to look at complexity. The successes have been successes of the small, where we could focus on isolated elements. In the United States, even though we spend more than any other country on healthcare, we have among the worst results among the industrial countries; certainly we are behind the Europeans and, in many ways, also behind Japan when the usual indicators of health are considered. This is something that worries public health people; why, they ask, do we spend so much and have so little to show for it compared to other countries?

Here are four hypotheses:

> a. One, we don't actually get more healthcare; we just spend more for it. We know that something like 20 per cent of our healthcare bill is in administration, that is, the cost of billing and the like. The rate of profit of the pharmaceutical industry is greater than that of capitalism as a whole, and much of that is in the United States. Doctors' salaries are huge, as are charges for hospital rooms. The consequence is that "investment" per patient is enormous.
>
> b. Two, even when we do get more healthcare, it is not always good healthcare. Now, this seems paradoxical because we have more MRIs and more CT scans and more dialysis machines than

most other countries. So why is our health not better? Medical decisions are not always made for medical reasons. There are a lot of incentives for making decisions as to which kind of techniques to use, what kinds of interventions — when to carry out heart surgery, for example — which give rise to differences in medical procedure among countries. We do a lot more implanting of pacemakers than Europe and perform more cesarean sections and hysterectomies. A hospital buys an expensive machine to attract both doctors and patients. But once on hand, it has to be used. You can't allow an MRI machine to sit idle in the hospital, so doctors are encouraged to use it if only to amortize the institution's investment. Another is that in order to keep the "batting average" of a surgeon high, he or she has to perform enough operations (several hundred a year) to keep skill levels up. An isolated hospital with only one heart transplant every three or four months is not a safe place to go. The wise patient will seek out a hospital with a highly regarded cardiac service equipped with the latest technology. But to win that prestige, skills must be maintained, so there's an incentive to keep both surgeons and machines working. Since the service is also an expensive thing to have, it needs to be kept busy if only to bring in surgical fees. But does it make sense to have all that expensive equipment? Hospital administrators will tell you it does because the hospital down the road has it. If Mass General is in competition with Beth Israel, and both compete with Mount Sinai (all three hospitals), all of them need the most advanced machines. Then there are the HMOs (Health Maintenance Organisation, medical-insurance conglomerates), which have their accountants making medical decisions, effectively rationing healthcare. Both approaches are meant to maximize profit. What happens is that sometimes people get too much care, sometimes too little. But in both cases, our health is a side effect of the obsession with making money. The irrationality of the system extends even to the rich, who are overtreated. We kill nearly two hundred thousand people a year through improper medical interventions. Many more die due to misuse of heavily advertised prescription medications, over-the-counter remedies, and other preparations.

c. The third hypothesis is that the U.S. healthcare system is built on a foundation of inequality. Only some of us actually receive or have access to the healthcare we need, while most don't.

d. Finally, the fourth hypothesis: we have created a sick society, even as we invest more and more to repair the damage. We are exposed to more pollution and increasing levels of stress and therefore exposed, ironically, to more opportunities to display our cardiac surgery skills. We make more people miserable, so we spend more on psychiatry and on psychotropic drugs. This is clearly evident in the public health situation in contemporary Russia, where the collapse of universal health coverage exposed the population to all the ills of incipient capitalism. They have had waves of epidemics, diphtheria, whooping cough, and the completely novel situation in modern times of declining life expectancy — from about 64 years to about 59 years. Ours is a sick society that demands ever greater expenditure to repair the damage to public health that it has itself inflicted.

RESPONSES TO THE CRISIS

The condition of healthcare has not gone unnoticed; in fact, there is widespread and growing dissatisfaction. And there have been a number of responses to address the situation.

Ecosystem Health

Ecologists looking at the problem have derived an approach they call Ecosystem Health. They posit that there are ecosystems under stress for multiple causes: from pollutants, contaminated food and water, high stress, and changes in the daily rhythm of life. For example, with nearly universal electrical light, people sleep less and our physiology changes. If we examine human biology as a socialized biology, we notice that there are things that appear as constants of human biology that really are not. For instance, it has long been conventional wisdom that, as a natural part of the aging process, blood pressure increases with age. But it turns out that among the !Kung Bushmen of the Kalahari, blood pressure increases with age only up through puberty and then levels off. Our blood pressure pattern is a function of the kind of society in which we live. We can see this in the pattern of stress reaction hormones that vary with one's social location. Recent Harvard studies have shown that among groups of teenagers from high school, all of whom are doing equally well academically, working-class kids showed prolonged rises in cortisol under the stress of exams while upper-class kids showed a quick spike and then decline. The physiology of working-class youngsters was altered by their social location, whether or not they acknowledged their working status. Evidently one's body knows one's class position no matter how well one has been taught to deny it. Human physiology, then, is a socialized physiology and differing social locations create different relationships with the environment. This knowledge has led to the ecosystem health concept, bringing together environmentalists and public health people to examine questions about how we rate the health of the whole ecosystem.

The Environmental Justice Movement

This movement arose from the observation — by others — that the best way to find an incinerator or a toxic waste dump is to look for an African-American neighbourhood. With lower real-estate values in minority neighbourhoods, it is cheaper to put the incinerator there. Zoning rules, made by the powerful, are more lax there. So the health risks from pollution and industrial waste become yet another facet of oppression. Exposure to pollutants doesn't affect everybody equally. Exposure to occupational health hazards — the exposure of somebody who makes a living sandblasting buildings, for example — is very different from someone who works at a desk totaling up actuarial tables. Exposure to environmental insult also varies with class and the condition of oppression. The environmental justice movement has been a response to this, fighting the dumping of pollutants and attempting to equalize the risks of an industrial society.

Social Determination of Health

This approach has been growing among epidemiologists, partly due to the rediscovery of what Rudolph Virchow and Frederick Engels pointed out in the nineteenth century: that capitalism itself can undermine health. This is important to keep in mind when conservative and reactionary commentators assert that there isn't any real poverty anymore. They argue that while some people make more money than others and can afford a bigger color television, the poor are not without their TVs. The car of a poor family is a little bit older or perhaps they do not eat in restaurants as often, but this inequality does not negate the real truth as the right-wing pundits see it: basically, there is no longer any poverty. Of course, an answer is easily found in the numerous studies that show that, in fact, black people pay for racist oppression with life spans ten years shorter than that of whites. Poor and oppressed minorities have 25 per cent fewer successful encounters with the healthcare system than more privileged groups. Meanwhile, the rate of death or other harmful outcomes increases with the level of poverty in illnesses like coronary heart disease, cancer of all forms, obesity, growth retardation in children, unplanned pregnancies, and maternal mortality.

Those interested in the social determination of health include some English scholars, such as Richard Wilkinson, who has looked at the life expectancies of different ranks in the English civil service. He found there was a difference even among those groups that are better off than those exposed to obvious dire need. He noticed that mere social hierarchy, social differentiation, makes your health worse everywhere, not only among those in extreme poverty. Now this can be interpreted in two opposing ways, but both of them are operative. One is to say that inequality *per se*, rather than the level of poverty, can make a person sick. Another is to say, quite literally, that it is all in your head. In support of the latter, baboon studies are cited that seem to indicate that those with higher rank in their troupe have better health. Their arteries are cleaner, they respond to stress like upper-class people; their cortisol level shoots up under stress and then comes right down again. Lower-ranking baboons tend to have the effects of stress lasting longer; their life expectancies are lower. But if you intervene in animal communities and alter their social hierarchy, within a few months the baboon's physiology will take on the characteristics of its new social location. This leads some people to say that it is how people perceive their situation in society — and therefore that people must be taught to cope with where they are, that after all we create our own realities. That's a common phrase in some of the growth and therapy movements: we create our own realities. It's not so much that you're underpaid and poor, but that you feel lousy about it. And so we have devised cheer-up pills: the cure for depression is not to get rid of the depressing situation, but to help people feel better about it. Another way to look at this so-called social determination of health is to see it not as a simple result of inadequate incomes that need to be raised, but as a consequence of a profoundly stratified, class-based society. Those who emphasize the latter feel that it is a more radical position than simply talking about how absolute deprivation is bad for your health, because the remedy for that would seem to be to increase income. Instead, they say, you have to eliminate the inequalities of class. Since the same studies can give rise to opposite conclusions, we need to emphasize that inequality affects your health in many different ways. When rich people think about poverty, they think about it only in the sense of having a little bit less, without examining the underlying structure of impoverishment. Poverty affects people, first of all, as chronic deprivation, actually having less food or worse food. Kids who live in damp, moldy apartments have worse health than kids who live in dry apartments. There are many other ways in which chronic deprivation itself is a menace to health.

There are what we call low-frequency, high-intensity threats, meaning those experiences that do not happen to everybody, but that could, and therefore are a constant threat to a sense of well being. Robert Fogel, a right wing, Chicago-school economist, pointed out in his book *Time on the Cross: the Economics of Negro Slavery* that most slaves were not whipped.[1] He went on to say that slavery was not what we would imagine from reading *Uncle Tom's Cabin*, that it had a certain economic rationality. What he neglects to say is that physical abuse of slaves, even when not employed, was a constant threat. Most slaves, perhaps, weren't whipped but all of them witnessed or knew about beatings. Similarly, most kids in impoverished neighbourhoods are not shot, but getting shot is a constant menace every time you go to the store or go outside. These are examples of medium- and low-frequency, but very high-intensity, threats.

There are also high frequency, low-intensity insults, the daily harassment one can see, for instance, in African-American communities. There, one is constantly forced to make strategic decisions. Am I walking so slowly that the cop is going to think I'm loitering? Or, am I walking so fast that he or she will think that I'm running away from the scene of a crime? If I come onto campus at night to work in my laboratory, will I first be stopped by the police who think that I'm a thief? I remember once the resident commissioner of Puerto Rico was stopped by police on the way to his office in Washington. They laughed when he said he was a member of Congress: the resident commissioner. Ramos Antonini was black.

We are learning now from the study of neurotransmitters that our brain is not the only locus of social experience. The cerebrum gathers social experience and transmits it through many branches of the nervous system into the neurotransmitters. The neurotransmitters are chemically similar to substances in our immune system, in the white blood cells. In a certain sense, we think with our whole bodies, we feel with our whole bodies, and so the whole body is the locus of social experience that comes with these patterns of chronic conditions, of low-frequency threats or high-frequency insult. There are many dimensions to the experience of deprivation but they are often lost in the hands of the statisticians, who simply see poverty as a quantitative difference in income.

The Health Care for All Movement

This group champions a national health insurance system and has done much work comparing the American system to the Canadian system; many progressive physicians are active in this movement.

Alternative Medicine

The alternative health movement deals mostly with individual health. It stresses diet, exercise, homeopathy, chiropractic, and naturopathic remedies — areas where people feel that they have not been treated adequately by the established medical system. They draw on a holistic approach to health rather than the targetted, magic-bullet approach of traditional allopathic medicine. They seem to be particularly effective in dealing with long-term chronic conditions rather than acute emergencies. For instance, for those who need radiation and chemotherapy for cancer, alternative practices are helpful in modulating the negative side-effects. The strategy of modern medicine is that cancerous tissue is sufficiently fragile and can, in effect, be poisoned in the hope that the radiation or chemotherapy will kill the cancer more than it kills you. The approach employed by alternative therapies not to attack the cancer directly, but to try to build up the body's defences. So the two approaches complement each other. Alternative medicine is very attractive and very powerful, but its primary appeal is to people who have control over their lives and access to the resources and techniques of alternative healthcare. It is not a mass movement; the holism it advocates stops at the edge of your skin. It is not a societal holism. Nonetheless, it is a powerful antidote to those movements that simply demand healthcare for all, without asking what kind of healthcare.

A RADICAL CRITIQUE

A radical critique of medicine has to deal with the things that make people sick and the kind and quality of healthcare people get. A Marxist approach to health would attempt to integrate the insights of ecosystem health, environmental justice, the social determination of health, "healthcare for all," and alternative medicine. One aspect of my approach to the issues of healthcare comes from my background as an ecologist. I looked at variability in health across geographic locations, occupational groups, age groups, or other socially defined categories. Just how variable, I asked, is the outcome in healthcare in different states in the United States, different counties in Kansas, different provinces in Cuba, different health districts in a Brazilian state, or in a Canadian province? Very interesting patterns emerged from that work. My colleagues and I examined the rate of infant mortality in each of these regions, both as an average and how, in each place, the rates varied, reflecting the quality of health-care, among other factors, from the best to the worst. What we saw was that infant mortality rates in United States were more or less comparable to Cuba, that Kansas had a rate a little higher than the U.S. average, while Rio Grande do Sur in Brazil had a more typical, and much higher, third-world infant mortality rate. That Cuba scored so high was not very surprising.

However, when we viewed the same data from the perspective of the range from the best to the worst rates of infant mortality, that is, the variability within given populations, an effective measure of fairness, much more was revealed. The numbers for counties in Kansas showed the greatest variation, while the numbers that compared U.S. states showed somewhat less difference. The difference across health districts of Rio Grande do Sur was even less, and the least variation was in Cuba. Similar things happen when we look at all causes of death. Once again, we observed average rates as well as the disparity; we divided the variation, the difference between best and worst, by the average. For Kansas the range divided by the average is .85, but in Cuba it was .34. We saw that the cancer rates in Kansas and in Cuba are comparable, but the variability is higher in Kansas than in Cuba. When we examined Canadian data, we found that Saskatchewan was somewhere between Kansas and Cuba.

The reason we chose these places is that on the one hand Brazil, Canada, and Kansas all have capitalist economies in which investment decisions are based on maximizing profit rather than any social imperative meant to equalize economic circumstance. Saskatchewan and Rio Grande do Sur along with Cuba have national health systems that provide fairly uniform coverage over a given geographic area. The Canadian and Brazilian regions have the advantage of a better and more just healthcare system than Kansas but, unlike Cuba, they have the disadvantages of capitalism, giving them an intermediate location in the variability of health outcomes.

This method can also be applied when comparing different diseases. One question we want to answer is whether variability will be greater across states and other large geographic regions, or across small areas like counties. There are good reasons why it might go either way. For example, weather could impact the data in large areas like states. But weather is not the only variable; others may vary greatly over smaller geographic units, only to be lost in the averages we develop for large areas. When we are able to look at smaller areas, for example, like different neighbourhoods within the city of Wichita, Kansas, we find a threefold variation in infant mortality. We also notice that unemployment in Kansas averages 9 or 10 per cent in most Kansas counties but is 30 per cent in northeast Wichita. Why? Because neighbourhoods are not simply random pieces of environment. They're structured. Wherever there is a rich neighbourhood, you need a poor neighbourhood, like northeast Wichita, to serve it. And so whenever we can get data across neighbourhoods, we see very large variations in social conditions and, as a consequence, in the quality and quantity of healthcare — clearly unnecessary from the point of view of any limitation in our medical knowledge or resources.

Another interesting case can be found in Mexico, where a study was conducted of several villages, ranking them according to how marginalized they were from Mexican life. Examined were such variables as whether there was running water or what proportion of the people spoke Spanish. The research showed that the more marginal communities had worse health outcomes.

But, unexpectedly, the data also showed that there was tremendous difference among the outcomes in poor villages that you didn't get among the villages that were integrated into the Mexican economy.

It is an as-yet-unrecognized ecological principle in public health that when a community or an individual organism is stressed for any reason (low income, a very severe climate, for example), it will be extremely sensitive to other disparities. So, if people have very low income, changing seasonal temperatures become very important. For example, in late autumn and early winter, emergency rooms have a lot of people coming in with burns from kerosene stoves, ovens, and other dangerous means used to compensate for inadequate heat in their houses. For such people, a small difference in temperature can have a big effect on their health — one that doesn't affect the more affluent. The same is true in relation to food. When people are unemployed, or if the prices go up, they cut back on food and other kinds of expenditures with an immediate impact on nutrition. If you are a superb shopper, and if you clip all the coupons and scrutinize the supermarket ads, you might just get by on the Department of Agriculture poverty level basket; the people who dream up these baskets assume you are a wiz at finding bargains. But suppose you are not so good, or that you read the ads but cannot get away for two hours for comparison-shopping. Or that you live in a neighbourhood where the local supermarket was not as profitable as the national chain that owned it thought it should be, and is gone, and with it your opportunity to get quality food. Or suppose that you would love to eat organic food for lunch but what you have is a half-hour break to go down to the vending machines. Under those circumstances, individual differences in where you work, how much energy you have, whether you can have a babysitter available or not, can have a big impact on your health.

THE ILLUSION OF CHOICE

Poor health tends to cluster in poor communities. Conservatives will say, "well, obviously poverty is not good for you, but after all, not all kids turn out badly. I made it, why can't you? Some people have become CEOs of corporations who came out of that neighbourhood." What they miss is the notion of increased vulnerability. The apparently trivial difference in experience can have a vast effect on the health of someone who is marginal. Suppose a pupil is a bit nearsighted but, because she is tall, is seated at the back of the classroom. The teacher is overworked and does not notice that the student cannot see the blackboard. She fidgets; she gets into a fight with the kid at the next desk. Suddenly she has become someone with a "learning problem" and is transferred to a vocational course even though she might have been a great poet. In a more affluent community, where the classes are smaller and teachers pay attention, this kid would simply end up with glasses. Individual differences can come from anything, from personal experiences growing up, even from genetics. But even when genetics is responsible for a given human characteristic, it is only responsible within a particular context. For instance, in a factory emitting toxic fumes, people will develop cancer at a higher rate; those most likely to develop the cancer have livers that are not able to effectively process a particular chemical as well. This is a genetic variable and thus a genetic disease but it occurs only with exposure to those fumes. The cancer is not a result of genetics alone; the environment also causes it.

Trivial biological differences can become the focus around which important life outcomes are located; the most obvious is pigmentation. The difference in melanin between Americans of African and European origin is, from the point of view of genetics and physiology, trivial. It is simply the way in which a pigment is deposited in the skin. Yet this difference can cost you ten years of life. So is this a lethal gene? Is this a gene for a higher spread of pigmentation — one that also makes you more vulnerable to arrest? A standard geneticist would look at family histories and determine that if your uncle were arrested, there would be a higher probability of you being arrested as well. Conclusion: the cause of criminality is genetic. Following the rules of genetics in this mechanistic way, he or she will have proved that crime is hereditary. This makes as much sense as the notion that black people get more tuberculosis because they have bad genes. Genetics is not an alternative explanation of social conditions; it is a component of an investigation of causal factors. There is an intimate interdependence among biological, genetic, environmental, and social factors.

Behaviour is one of the areas where public health workers want to intervene, arguing that much that differentiates health outcomes in poor neighbourhoods from rich ones can be associated with behaviour, such as smoking, exercise, and diet. Conservatives, finally forced to concede that there are big differences in health outcomes between rich and poor, now say, "Yes, this is because the poor make unwise decisions. The appropriate remedy is education. We know that kids do better if their mothers have had more schooling, so what we need are education programmes to teach people to make the best of their situation." In fact, some health education programmes are valuable. Safety orientation within factories does help people cope with unsafe conditions. But let us take a closer look at this question of choice. The Centres for Disease Control, and others who deal with these issues, say only some things can be chosen, while others are imposed by the environment. They would have us distinguish between disadvantages imposed on us, that may be unfair and/or can be eliminated, from those that were freely chosen and for which we can only blame ourselves. A Marxist confronted with choices among mutually exclusive categories like choice versus environment, heredity versus experience, biological versus social, knows that the categories themselves must be challenged. Choice also implies the lack of choice. Choices are always made from a set of alternatives that are presented to you by somebody else. We know this from elections and from shopping. We choose food, but only from the products a company has chosen to make available to us. The choice is distinguished by the lack of choice, that is, unchoice. The same is true with respect to the opportunity to exercise choice. There are always preconditions to the exercise of choice. If the conditions of life are very poor or oppressive, some of the things that are unwise choices under other circumstances become the lesser evil.

Public health people, like nearly everyone else, worry a great deal about teen pregnancies, which generally are not a good idea. Teen mothers are not experienced; they may have difficulty taking care of their babies; and the babies are more likely to be underweight. Nevertheless, it turns out that the health of a baby born to an African-American teenager is on the average better than the health of a baby born to an African-American woman in her twenties. Why? The environment of racism erodes health to such an extent that it makes a certain amount of sense to have your babies early if you're going to have them. This is something that is not obvious when you simply say, "teen pregnancy is a danger to people." We need to look at teen pregnancy in a much broader social context before we can think about making it simply a public health issue.

Smoking is another example. Smoking increases inversely with the degree of freedom one has at work. People who have few choices in life at least can make the choice to smoke. It is one of the few legitimate ways in some jobs to take a break and step outside. So there are people who choose: "yes," they say, "it might give me cancer in twenty years, but it sure keeps me alive today." The unhealthy choices people make are not irrational choices. We have to see them as constrained rationality, making the best of a bad situation. Most of the apparently unwise decisions people make have a relative rationality to them when their circumstance is taken into account, so it is unlikely their behaviour will change simply by lecturing to them. You have to change the context within which choice is made.

Yet another dimension of choice is found in the way we perceive time. When making a choice about health, we assume that something we do now will have an impact later on. That may seem obvious, but it is not the experience of everyone. Most people, in fact, do not experience the kind or quality of freedom that gives them control over their own lives, that would allow them to say, "I will quit smoking now so that I won't get cancer in twenty years." Not everyone can organize his or her lives along an orderly annual time scale. In the inner city of San Juan, in Puerto Rico, the life pattern is such that one can work unloading a ship for twenty-three hours a day for two days, then sleep for three days, then unexpectedly work in a restaurant for another two days because his or her cousin has to go to a funeral in the mountains. Time does not have the same structure when you can't make solid plans now for what is going to happen to you later.

On the other hand, the lives of, say, academics are notable for the way time is organized. Students can and do choose courses of study that, in two or three years, will prepare them for a career. On a shorter time scale, a professor may conveniently order his or her teaching schedule around patterns of Monday, Wednesday, and Friday, or Tuesday and Thursday. Physicians decide when to see patients, when to be in the library, when to go to seminars. So some people can actually structure their lives in such a way that we can actually make predictions. Not absolute predictions, obviously. Things can come up; a car can hit us. But, basically, the more control you have over your life and your experience of life, the more it makes sense to make the kind of decisions that public health experts recommend, the more the possibility, then, of exercising choice. So the answer to those who talk about decision-making and choice is to tell them, first of all, to expand the range of choices. Secondly, they need to provide the tools for making those choices. Third, of course, people need to control their own lives, so that they can exercise all their faculties to make meaningful choices. In taking each of these steps, we directly challenge the false dichotomies that rule thinking about public health and constrain it within predetermined societal boundaries.

WHAT CAN BE DONE?

At a recent meeting I attended, a paper was distributed that posed the following dilemma: Why, living in a democracy, where all citizens have the vote, do we permit policies that create inequalities that have such a negative impact on our health? How do we explain this? We have schemes to improve agriculture but they increase hunger. We create hospitals and they become the centers for the propagation of new diseases. We invest in engineering projects to control floods and they increase flood damage. What has gone wrong? One answer might be that we are just not smart enough. Or the problems are just too complicated, or we are selfish, or we have some defect. Or, after having failed to eliminate hunger, improve people's health, and do away with inequalities, and failed, perhaps we need to face facts and conclude that it just cannot be done. Or perhaps we're just the kind of species that is incapable of living a cooperative life in a sensible relation to nature.

We should reject any of these unduly pessimistic

conclusions. The history of struggle is long and not without achievements. But struggle is also difficult. For example, it is easy to depend on the illusion of democracy and a beneficent government to solve our problems. But when we look at the policies that emerge from those institutions of democracy, we see that those ostensibly aimed at improving the people's lives are nearly always hobbled by some hidden side condition. For instance, I am sure that on the whole, most presidents of the United States would rather have people covered by health insurance than not. But that is subject to the side condition that insurance industry profitability must be protected. They probably would like medicines to be cheaper, but only if the pharmaceutical industry continues to make high profits. Abroad, the United States would like peasants to have land, but only if not expropriated from plantation owners. The basic reason that programs fail is not incompetence, ignorance, or stupidity, but because they are constrained by the interests of the powerful. Sometimes we discover that part of a programme is carded out successfully, and part not. An enterprise zone might be established in an inner city that actually brings in investment, but there is no impact on poverty because the assumption that benefits would trickle down was an illusion. A reasonable return on investment was the goal of the developers. When that was achieved, nothing else mattered.

A good way to see how these hidden constraints, these systemic barriers, operate is in the delivery of health services elsewhere. Healthcare in the United States exists against the background of this country's unrestrained capitalism. We have described at length both the prospects and problems of that system. But, in Europe, social democrats historically have taken a different approach — one that acknowledges inequality as an obstacle. They have treated unemployment, for example, as a social problem rather than an inevitable byproduct of a vigorous market. A town council will address it by financing a centre for the unemployed, with counsellors to advise them of their right to unemployment insurance and other benefit programmes. The centre may even organize a support group where people can deal with their feelings about not being able to bring home an income to the family. Local governments can address other social concerns. In London, there is a programme to break down the isolation of young mothers, where they can meet one another, share experiences, and provide support. Of course, none of these measures affects profitability or challenges the market. So the council cannot create employment. Even the most farsighted programmes initiated by European social democratic governments do not challenge the capitalist order in any way. What they do is to try to make things more equitable — for instance, through progressive income taxes or generous unemployment insurance. In Sweden, transport workers demanded improved food to reduce heart disease among truck drivers. They organized to improve the quality of food in the roadside canteens and collaborated with restaurant owners and canteen owners and food was improved. In other places, unions have negotiated collective agreements to change shift work, hours of work, and working conditions. The unions recognized that health concerns were but another aspect of class relations.

In some cases improving on-the-job health is relatively cost-free. No employer will object to putting up a sign reminding workers to wear their hard hats on the construction site. But it begins to get a little tricky when you talk about the reorganization of work or the expenditure of money. If the expenditure of money comes from taxes, through government programmes to improve health, we can expect the business class to object. And if, after each new expenditure, they perceive some interference with their competitive position, their opposition may take some political form, for example, the repeal of some aspect of health and safety regulation. When an expenditure has to come from the individual employer, perhaps byway of a union demand, they will be even more resistant. They will say that it is bad for competition and threaten to close down and move somewhere else. If the union's demands deal with the organization of work itself, management will see workers impinging on the very core of class prerogative. In that situation, only a powerful and well-organized labour movement will be able to impose changes.

When health policy is looked at from the point of view of which issues involve a direct confrontation of fundamental, ruling-class interest, which ones involve simply relative benefits to a class, and which are relatively neutral, we can predict which kinds of measures are possible. This highlights the lie in the notion that society is trying to improve health for everybody. We need to see healthcare in a more complex way. Health is part of the wage goods of a society, part of the value of labour power, and therefore a regular object of contention in class struggle. But health is also a consumer good, particularly for the affluent, who can buy improvements in health for themselves. Rather than improve water quality, they buy bottled water; rather than improve air quality, they employ oxygen tanks in their living rooms. Health is also a commodity invested in by the health industries, including, hospitals, HMOs, and pharmaceutical companies. They sell healthcare to as large a market as can afford to pay for it; they even push it on people who do not need it. Like any aggressive business, the health industries engage in public relations — the winning of hearts and minds. Some of the clinics that were established in Southeast Asia during the Vietnam War and earlier, during the Malayan insurrection, were for this purpose. Doctors, at great sacrifice, would go into the jungle and set up clinics and work very hard under very difficult conditions for low pay, seeing themselves either as bringing benefits to people who needed it or, more consciously, as trying to prevent communism. It was yet another reincarnation of the White Man's Burden that justified nineteenth-century imperialism.

If good health depends on one's capacity to carry out those activities that are necessary and appropriate according to one's station in life, it matters how that station is determined. Those who can determine for themselves what constitutes necessary and desirable activities are clearly different from the people who have that determination made for them. This distinction is clear when an employer negotiates health insurance for his or her employees; for the employer, the cost of the benefits package will always come before what employees may think they need. So health is always a point of contention in class struggle. So is medical and scientific research; knowledge and ignorance are determined, as in all scientific research, by who owns the research industry, who commands the production of knowledge. There is class struggle in the debates around what kind of research ought to be done. Increasingly, research in the health field is dominated by the pharmaceutical and electronic industries.

There are intellectual concerns about how to analyze data, about how to think about disease, about how widely we need to look at the epidemiological, historical, and social questions they raise; there are also issues of health service and health policy. But they are all part of one integral system that has to be our battleground in the future. We have to take up health as a pervasive issue as we do with problems of the environment; they are aspects of class struggle, not an alternative to it.

Six
Cuba's Fidelity to the Earth

The question I will try to answer is: how did Cuba do it?

While the world environmental problems continue to worsen despite intensive research and rhetoric, how come a poor Third World country besieged by a hostile neighbour has been able to embark on an ecological pathway of development that combines sustainability, equity, and quality of life goals? How did it achieve a commitment to an integral programme of protected areas, ecological and organic agriculture, public health levels behind only the Scandinavian countries, environmental education, occupational health, urban planning and economic development compatible with environmental protection, and compliance with the major world treaties on the environment?[1]

Although the commitment to agroecology and ecological development is relatively new, it is not, as is often misrepresented, an improvised emergency response to the Special Period, the economic crisis brought on by the collapse of Cuba's trade relations with the Soviet Union and the tightening of the United States' economic warfare. Rather, it has its roots in a complex history of colonial science, anti-imperialism, the emergence of a self-conscious community of ecologists, and the transformations of Cuban society since 1959.

Scholars of different disciplines tend to prefer different analytic frameworks, to explain observed processes. Historians may trace the sequence of steps from the early literacy campaigns and plans for botanical gardens through the present national environmental plan. Sociologists might point to the institutional frameworks, the role of the U.N., government departments, research institutions and NGOs that analyze and propose.

Policy analysts might focus on the legal framework and when and how particular decisions were made, or look at the unique events, the right person in the right place at the right time to achieve the ecological commitment. Intellectual historians might show the unfolding and deepening of environmental awareness and concern, the conflicts about pesticides, and the philosophical underpinnings that made the outcome almost inevitable. Economists might point to the Special Period and show how the urgencies of scarcity forced a rethinking of industrial and agricultural strategy.[2]

As a Marxist, I view these approaches as different modes of abstraction applied to the same complex, multilevel reality, the whole that is the full explanation. Therefore, I will try to place these various descriptions and interpretations in the context of an evolving Cuban socialism.[3] A complex, nuanced explanation of a phenomenon is not the antithesis of theory and generalization. Rather, it demands a theory of complexity and process.[4]

My primary concern is not a description of the state of the Cuban environment or a catalogue of successes and failures but rather the trajectory of Cuban society's evolving relation with the rest of nature. My thesis is that each kind of society develops its own relations with nature, and that an ecological pathway of development is at least latent in socialist development, co-equal with equity and participation. Despite all the zigzags, vacillations and disputes, it emerges as an increasingly central characteristic. And this is imperative, for socialism cannot succeed without committing to an ecological pathway. Indeed, the failure of the Soviet Union and Eastern Europe to do so was one symptom of the disintegration of the European socialist project.

I will start with Cuban science in general, then environmental science and policy, and finally the specific case of agriculture.

SCIENCE

José Marti's modernist valuing of learning[5] was joined with the traditional socialist appreciation of science to encourage the young revolutionaries to give a high priority to science from the earliest days of the revolution. The traditional socialist view was that scientific knowledge had been produced out of the wealth created by working people but was monopolized by the rich to be used for profit and to build the instruments of power. Therefore the recapture of scientific knowledge for the people was a common goal of radicals throughout the world, and any scientific learning was considered a victory. Further, scientific literacy was seen as liberation from religious obscurantism and bigotry. Scientific news or controversies frequently appeared in the socialist and communist publications. Public lectures in England, the United States, and Russia contributed to this goal. My own grandfather, who had a third grade education, believed that every socialist worker should at least know cosmology, history and evolution. In pre-revolutionary Cuba the *lectores* (readers) in the tobacco factories were hired by the workers to read from world classics and scientific literature while they worked.

Thus it was natural for Cuban revolutionaries to look toward science for economic development and as part of the necessary culture for a free people. In 1960, Fidel Castro was invited to speak to the Cuban Speleological Society.[6] In that talk he proposed that

"The future of our country will be a future of men of science." (In Silvia Martinez' 2003 book, she corrects the sexism and paraphrases it as "men and women of science.")[7]

The preconditions for today's modern Cuban science were laid in the early years of the revolution with the literacy campaigns, starting with the battle for the sixth grade. The enemies of the revolution realized the significance of education, and CIA-sponsored counter-revolutionary bands murdered two young alphabetizers, Conrado Benitez Garcia and Manuel Ascunce Domenech. But the country became fully literate and has continued to extend mass education through secondary and, increasingly, university level, with university centres in every municipality and special programmes for seniors, dropouts, people with disabilities, and workers displaced by the downsizing of the sugar industry. Now Cuba, with only 2 per cent of the population of Latin America, has 11 per cent of its scientists, a large fraction of them women. More than 1.3 per cent of the population works in science, a level comparable to the most developed countries. There are more than 100 major research centres as well as research sections of institutions dedicated to other goals. Physics alone has 40 laboratory groups with some 500 researchers concentrated in solid state and nuclear physics, optics, space and geophysics, mathematical physics and medical physics. There are some 80 centres doing social science research, studying such topics as marginality, social dysfunction, issues of race and gender, and residual and newly arisen inequality. Cuban science has been outstanding in the areas of public health and medicine, agriculture, electronics and pedagogy. Table 1 (below) shows some of the major achievements of Cuban science.

Table 1. Some Major Achievements of Cuban Science and Technology

- 271 new medications.
- 24 diagnostic systems.
- SUMA (ultramicroanalytic system for HIV detection).
- Production of 90 per cent of the medicines needed.
- Melagenina (84 per cent effective against vitiligo).
- Meningococcus B vaccine.
- Hepatitis B vaccine.
- Haemphilus influenzae vaccine.
- Skin growth factor for treatment of burns.
- Monoclonal antibody HB3 for epithelial tumors, especially of the head and neck.
- PPG anti-cholesterol agent.
- Anti-retroviral medications against AIDS.
- Control of HIV/AIDS to .03 per cent of at-risk population through detection, quarantine, treatment and education.
- Infant mortality at around 6.5 per 1000 live births (tied with Canada for best in hemisphere).
- Elimination of poliomyelitis, malaria, and infant AIDS.
- Integrated programme for the treatment of *Retinosa pigmentosa*.
- Neuro-rehabilitation Centre.
- Orthopedics: development of external fixators.
- Psychiatry: emphasis on outpatient care, occupational therapy, and integration into the community.
- 100 per cent HIV-free blood bank.
- Sanitary seed production (*vitroplantas*).
- Sugar cane derivatives for coolants, medicines, energy, and paper production.
- Biological methods for preserving and enhancing soil fertility.
- Systems of biological control of pests through release of parasites.
- Reforestation and protected areas.

For any Third World country the problem arises, how can it create a science, which is at the same time international in the sense of being linked to the advanced communities of world science, and yet have its own agenda directed at the needs of its society? And how can it do this on a low budget? In Cuba this comes to some 10,000 pesos per scientific worker. The peso is currently 26 to the U.S. dollar, but this grossly underestimates its real value within Cuba. In the U.S., the expenditure is about $200,000 per scientist.

The lack of resources make very strict prioritizing a necessity. This produces a glaring unevenness of availability. While everybody who needs kidney dialysis receives it, reagents are often not available for student chemistry laboratory exercises. As of this writing, a group of my colleagues studying mosquito development in breeding sites in Havana are unable to measure the water temperature regularly, because they do not have sufficient thermometers.

Cuban science is right up there with the best of world science, but it also has its own special features. First, it is publicly owned science. With the recent programme to establish university centres in all 149 municipalities and to make all centres of learning "micro-universities," research is even more broadly diffused.

Public ownership makes it possible to plan science, to incorporate science into national plans, and to have policies linking the recruitment and training of scientists to research directions. The mass literacy campaigns of the 1960s enlarged the pool of potential scientists, while the struggles for women's equality and against racism opened up new sources of talent. Today, women make up 52 per cent of the scientific work force[8] and 65 per cent of the scientific and technical workforce,[9] many in positions of leadership. Among the people I have worked with, the Minister of Science, Technology and the Environment is a woman. The dean of the mathematics faculty at the University of Havana is a woman. So was the director of the Institute of Fruit Research, the head of the Centre for Animal and Plant Protection, its laboratory of phytopathology, the Centre for Mathematics and Cybernetics Applied to Medicine, the Carlos Finlay Institute, and two of the four department heads at the Institute of Ecology and Systematics. A growing number of the scientists are Afro-Cubans. Education is free of charge. Study is regarded as productive labour, the task of producing a skilled and well-informed citizen, rather than as an investment in future high income. Therefore, there are no economic barriers to study.

When the British Marxist, J.D. Bernal, first posed the need for the planning and collective scientific work in the 1930s,[10] the very idea was met with derision and hostility. It was denounced as the totalitarian repression of free, individualistic science. But now scientific strategies are an accepted part of government policy throughout the world.

Scientific planning confronts a major contradiction: how do you plan for the still unknown? We can assess national needs and establish priorities. But surprise is inevitable in science, and therefore it is necessary to be able to follow new pathways that present themselves during the lifetime of the plan. But how do you recognize surprises that should be followed up? Exciting new directions and innovations are not yet the consensus of a scientific community, but rather the initiative of a few individuals. On the other hand, the agreed upon priorities represent the consensus of the leaders of science — those who created the field the way it is — and therefore, who are less likely to be critical of its direction. This makes it necessary to have a very flexible form of planning with leeway for departures from the plan.

Cuban scientific planning sets general goals. At a national level, some general research goals are proposed as priority areas.

Provincial governments have their own priorities, and so do the various ministries and institutions. The individual institutions join in the projects that fall within their areas, and there is leeway for individuals to carry on their own work where resources permit. The projects that come down from above have actually gotten there at the initiative of the researchers so that the process of planning moves up and down many times in the formal structure before a plan is adopted.

There are frequent progress reports during a scientific investigation. This includes discussions with peers and the public that will be affected so that the research is an even more collective process than the interdisciplinary teams would suggest.[11] In 2001, 3,093 formal research results were reported, of which 403 came from national programmes, 1584 from ministries and branches of the society, and 1,077 from territorial authorities. From the non-professional end, the National Association of Innovators and Rationalizers (ANIR) presents some tens of thousands of innovations each year, which indicates the breadth of mass participation in Cuban innovation. (ANIR has more than half a million members who have come up with over 100,000 solutions to mostly technical problems.) This experience refutes the notion that innovation would stagnate in the absence of opportunities to get rich from inventions. Amateur groups in computing, botany, and other fields supplement the professional research.

The public nature of science makes it open science. There is no hiding of information for proprietary reasons, as is increasingly common in the U.S. and other capitalist societies. Inventors can receive economic rewards from their inventions, but they do not have the authority to suppress them or restrict their use. There is little duplication of effort by competing entities. This makes it possible for Cuban scientists to collaborate across institutional boundaries. The development of the recent completely synthetic vaccine against *Hemophilus influenzae* was the work of many research centres concerned with the biochemistry, immunology, clinical and industrial aspects of this innovative project. The National Study of Biodiversity was prepared through the collaboration of the Ministry of Science, Technology and the Environment (CITMA), Ministry of Higher Education, Ministry of Agriculture, Ministry of Public Health, Ministry of the Economy and Planning, and many institutes within these ministries. The same was true for the National Atlas and other major efforts; all involved broad collaboration. In the absence of the obscene race for patents, Cuba has the capacity to wait and see what unexpected consequences an innovation may have. Cubans have been working with genetically modified organisms for more than 17 years but have not released any GMO plant varieties, because they are still exploring the possible risks to the environment.[12]

A major obstacle to the serious evaluation of research and policy efforts in the U.S. is that the annual reports of grant holders and agencies to their sponsors are a mixture of real assessment and self praise, a pitch for more funding that downplays difficulties. The large scale of many projects makes it impossible to replicate or for outside examiners to know them well enough to evaluate critically. CITMA is able to look more objectively at the state of the environment and identify weaknesses. In January 1997, CITMA convened a workshop on the environment as a national consultation, "Rio.5," to evaluate Cuban compliance with the agreements[13] that came from the U.N. Conference on Environment and Development, the Earth Summit that had been held in Rio de Janeiro in 1992.

Invitations were extended to various government departments, agencies, NGOs and delegations from nearby provinces. Participants considered each of the categories discussed in Agenda 21, the declaration and plan of action that emerged from the Rio Earth Summit, and also added several of their own. For each problem area, they described achievements and also the major difficulties. For instance, in the chapter on agriculture, the list of achievements includes the system for plant disease prediction, the establishment of centres for the reproduction of natural enemies of pests and diseases, and the expansion of polyculture. Among the deficiencies listed are insufficient and unstable extension work, lack of studies of the environmental impacts of new production systems, and, especially interesting, "the existence at various levels of the opinion that the practice of sustainable agriculture is only a consequence of the Special Period destined to disappear when the present limitations make it [possible] and there will be a regression to high inputs of fertilizers, pesticides, mechanization, etc."[14]

Later that year, The National Environmental Strategy evaluated recent experience (up to 1997). After listing their achievements, they report (my editing of the Cuban version in English):

> Parallel to these gains there have been mistakes and shortcomings, due mainly to insufficient environmental awareness, knowledge and education, the lack of better management, limited introduction and generalization of scientific and technological achievements, still insufficient incorporation of the environmental dimension in the policies, developmental plans and programmes, and the absence of a sufficiently integrative and coherent juridical system. Moreover, the scarcity of material and financial resources has prevented us from attaining higher levels of environmental protection, which has worsened in the last few years due to the economic situation in which the country has been immersed because of the loss of commercial ties with the former socialist bloc and the continued and intensified economic blockade by the United States of America.

Although it is often asserted in the United States that the Cuban government blames the United States for all its troubles, here we see a nuanced analysis that includes the economic warfare against Cuba as only one factor among many that influence the situation.

One common misconception is that the science of a developing country should concentrate on the applications of the achievements of world science and that fundamental research is a luxury of the rich. However this condemns a country to dependence on the basic research done in other countries for other reasons. A coherent scientific community has to develop its own underpinnings for its work and the education and morale of its participants. In the National Environmental and Development Programme, established by CITMA in 1995, there is a group working on protection of the atmosphere. It includes such immediately practical tasks as the monitoring of air pollution on different temporal and spatial scales and linking this to morbidity and mortality data. It studies the chemistry of rain and the ocean/atmosphere exchange. The group also includes neurobiologists who work with problems of autism, trauma and the neurological correlates of emotion, investigating potential links between air pollution and these issues.

In agriculture, a doctoral thesis has to include a section on the contribution of the work to practice and how it enriches science. However, Cuban science includes themes that are not directly related to practice, such as underwater archeology and the theory of complex systems. For example, a recent international symposium[15] on complexity included presentations by Cubans such as: Entropy and Complexity: the Problem of Irreversibility; Contingency and Causality in Natural Disasters; Complexity and Morphogenesis: From the Properties of Systems to the Very Existence of Systems; Construction of a Critical-Analytical Model for the Study of Cultural Identities in the Social

Complexity; Aesthetics and Reasoning about Complexity: an Epistemological and Methodological Approach; Evidence for the Mind as a Dynamic Attractor; The Treatment of Attention Deficit as a Non-equilibrium State; Transformations of a Citrus Agro-ecosystem in Conversion to Organic; and The Complexity Barrier: the Next Challenge for Immunology. As is increasingly common in Cuba, the symposium included musical performances as part of the plenary sessions.

There is a special style to Cuban science that is very much influenced by the Marxist dialectical philosophy of science, with its emphasis on historicity, social determination of science, wholeness, connectedness, integrated levels of phenomena, and prioritizing of processes over things. All doctoral candidates must study Social Problems of Science and Technology, which has emerged in the 1990s as a distinct field of study. The impact of this preparation can be observed in the self-conscious view of the development of science as a social process, with the organization, recruitment, priorities, preferred approaches and tools of investigation all being recognized as products of the social relations that promote, support, apply and reward scientific endeavor. This allows for a critical examination of the state of a field internationally and the capacity to make active choices about what to concentrate on. For instance, by recognizing that the pharmaceutical industries develop only those medications for which there are large and lucrative markets, the Cubans have been able to select areas of research that are ignored, because the knowledge is not easily turned into commodities, or because the disease is uncommon among the wealthy. Thus Cuba has been in the forefront in work on Retinosa pigmentosa, vitiligo, and malaria. It also encourages a view of science that combines its contributions to the economy and to the general culture of society with an awareness of its own internal needs for balance among the disciplines, the integration of practical and theoretical concerns, and the cooperative organization of research.

A major characteristic in the Marxist dialectical perspective is wholeness and the critique of reductionism. A recurrent theme in all of Cuban science is the breadth with which problems are approached and the willingness to span levels of organization. Agostín Lage, immunologist and director of the Centre for Molecular Immunology, has been an outspoken critic of molecular and genetic reductionism. He sees the immune system as "a system of recognition and control of the composition of its own organism, whose regulation depends not only on the presence or absence of specific cellular clones but also on the interaction of these clones among themselves (supra-clonal properties)."[16] He sees the future of immunology as including interaction with neurobiology and calls for a synthesis of the high-tech molecular sciences with social medicine. Lage also raises ethical issues in science, particularly the question of whether science is used to increase or remove inequality in the world.

This multilevel approach pervades much of Cuban science. In medicine, modern technical tools coexist with herbal medicine (the "green pharmacy"), social epidemiology, and various kinds of alternative medicine. These are not seen as in opposition: the Carlos Finlay Institute, which has pioneered in the development of molecular biology to produce vaccines, antibiotics and anti-cholesterol agents, also has a controlled experimental programme for testing the macrobiotic diet. The very successful Cuban response to HIV/AIDS has combined chemotherapy (everyone who needs retroviral drugs receives them) with population-level interventions that include temporary quarantine and community education. Work in rehabilitation combines advanced neuroscience and occupational therapy. The prevalence of HIV/AIDS in Cuba is currently about 0.035 per cent, and there have been no cases of infant infection since 1997. As a result of its strong commitment and broad approach, Cuba is the healthiest Third World country and is tied with Canada for the lowest infant mortality in the hemisphere, making it a global health leader.

Another dialectical theme is the priority given to processes over things. Nilda Perez, a leading Cuban agricultural ecologist, poses the direction of change as from an agriculture of inputs to an agriculture of processes.

Science is defined more broadly in Cuba than in the U.S. The recent colloquium that produced the book, *Cuba, Dawn of the Third Millennium: Science, Society and Technology*, included participants from the usual fields as well as economics; pedagogy; science, society and technology; and communications and the audiovisual media. The discussions started with each participant describing the state of his or her own field and perspectives for the future and then entered into open discussion around general questions. Ethical issues were recurrent themes.

Thus Cuban science made possible the effective commitment to an ecological pathway of development by being publicly owned, planned, collaborative, holistic, multilevel, integral to the education of all Cubans, and committed to meeting the material and cultural needs of the people.

The above description stresses the directions in which Cuban science is different from science in capitalist societies. In Cuba, not all institutions work the way they are supposed to, not everybody thinks dialectically, and we can always find examples of narrowness and parochial interest. But the significance in Cuba is the pathway that is being built, the direction of change.

DEVELOPMENT OF AN ENVIRONMENTAL PROGRAMME

The political programme of the Cuban Revolution did not have an explicitly ecological perspective at the beginning. The urgent concerns of the new government were eliminating extreme poverty, providing water and sanitation, housing, and literacy. But even before the triumph, the commander of the 26th of July Movement in Matanzas, Onaney Muñíz, was planninq for a botanical garden. The destruction of Cuba's forests, the erosion caused by monoculture and the sugarcane economy, the prevalence of infectious diseases that could be prevented, and the need to develop the resources of the country to eliminate poverty all led to the creation of separate programmes that later nourished ecological development as a conscious goal.

The literacy campaign of the Sixties made possible the massive commitment to science, which in turn laid the foundations for environmental sciences. As soon as Rachel Carson's *Silent Spring* reached Cuba, Fidel Castro was circulating it among his associates, and environmental consciousness began to spread. Already in the 1960s there were programs for reforestation, introduction of the Voisin system of rotational grazing, the digging of thousands of ponds as micro-reservoirs, cleaning up of foci of infection, and massive immunization campaigns. The Institute for Physical Planning, a new discipline for Cuba, undertook the first environmental studies for the selection of sites for development.[17] El Grupo, the Centre for the Integral Development of the Capital, is one of the catalysts for innovative participatory neighbourhood development.

The priority goals in agriculture were a stable food supply, income and safety for the rural population, sugar for export, and inputs for industry. The dangers in the use of pesticides were recognized and confronted, at first, mostly by measures to protect agricultural workers. But agricultural development was still largely within the paradigm of the "Green Revolution," which depended on high yielding plant varieties and massive infusions of mechanical and chemical inputs, most of which had to be imported. There was not yet any organized field of ecology. Cuban biology was typical of colonial biology throughout the tropics: biology applied to medicine and agriculture, and systematic botany and zoology, with systematists making ecological observations. At the University of Havana the zoology curriculum started with two years of mostly descriptive zoology, surveying the major families of animal life. In discussions within the university in 1968, the administration and the students supported proposals for electives in ecological topics, but many in the faculty thought this would force the omission of whole families of animals from the curriculum, "thus losing the whole picture of evolution," as if evolution were a catalog of its results. In any case, the faculty was not yet itself prepared to teach those subjects. But there was already experimentation with Voisin's rotational grazing system, polyculture, and biological pest control.

Table 2. Some Milestones in the Development of an Ecological Pathway

1960s

- Founding of the Institute of Physical Planning and the Group for the Integrated Development of the Capital.
- Introduction of the Voisin system of rotational grazing.
- The beginning of restoration of open pit mining areas.
- Construction of some 1,400 micro reservoirs for energy, water resources, recreation, and fish production.

1970s

- Transition toward low-input agriculture.
- The creation of botanical gardens.

1972–73

- National Atlas.

1974

- Cuba joins UNESCO's Man and the Biosphere programme and selects the Sierra del Rosario montane rainforest as its study area.

1975

- Zoning of Havana.
- First Congress of the Communist Party adopts thesis on environment.
- Waste treatment facilities are required for all new plants.

1976

- Constitution adopted. Article 27 links environmental protection to sustainable economic and social development and recognizes the obligations of the state and citizens to protect the environment.
- COMARNA (Commission on Natural Resources and the Environment) established.

1978

- Law passes allowing veto of development projects that would harm the environment.

1980s

- Experiments with ecological agriculture and *organopónicos*.
- Establishment of the Centres for the Reproduction of Entomoparasites and Entomopathogens (CREE) for biological pest control.
- Legal and institutional structures for environmental inspection and licensing of development projects are created.
- Plan Turquino-Manatí for the sustainable development of the mountains is implemented.
- Widespread adoption of urban *organopónicos*.

1990s

- National Environmental Strategy is implemented.

- Legal instruments for environmental protection, inspection and enforcement are developed.
- National Survey of Biodiversity is established.
- Antonio Nuñez Jimenez Foundation for Nature and Humanity vision statement: "a Cuban society with a developed environmental consciousness that recognizes nature as part of its identity, and an institution active in the development of environmental and cultural values in Cuba and the world."
- Network of protected areas established.
- Forest cover reaches 23 per cent of land area.
- System of Protected Areas.

Table 2 shows some major events in the evolution of an environmental perspective and commitment. In general, the 1960s were a period of laying the foundations for later development. The public health system was able to eliminate polio by 1963, malaria by 1968, and diphtheria in 1971. The first law of agrarian reform in 1959 made national land available for development programs. Within three years, illiteracy was almost eradicated. The abolition of legal racism, the recognition of equal rights for women, and the expansion of free education and scholarships widened the pool of potential scientists. Cuba sent thousands of students to study abroad, mostly in the east European countries. Maps were prepared of soil, water resources, and endangered species. In the 1970s, Carlos Rafael Rodriguez introduced his argument differentiating development from growth and arguing for integral development, laying the groundwork for a goal of harmonious development of the economy and social relations with nature. This implied a rejection of Stalin's approach, the view popular among east European communist governments that production decided everything, and that only after abundance was achieved would society be able to confront the task of bringing social relations into harmony with the economy. Despite his disagreements with Rodriguez on how the economy should be organized, until he left for Bolivia in 1967, Ché Guevara had already been stressing that social relations and economic development must evolve together. At the same time, UNESCO initiated its International Biological Programme, a ten-year international programme of biological studies that concentrated on the productivity of biological resources and human adaptation to environmental change. Cuba joined in and selected the montane rainforest of the Sierra del Rosario as its area of study. There, sitting on the sopping litter under an endless tropical rain, zoologists and botanists came together to begin to think of themselves as ecologists.

At the first national meeting on ecology in 1981, which I attended with the plant ecologists, representatives of research groups in botany, zoology, agriculture, oceanography, and from the tourism and food processing industries gathered to debate pesticides and consider what can be done with industrial wastes. The food processors called our attention to the pollution they were causing and asked what they could do with the mountains of rice husks and mango pits they were accumulating. The tourism institute asked how to develop environmentally friendly facilities. We ended the meeting with a resolution calling for the Commission on the Environment to have enforcement powers. This was soon accomplished, and the Commission was elevated to cabinet status, the present Ministry of Science, Technology and the Environment (CITMA). Its formation "resolved a contradiction in the old structure of leadership of environmental activity in the country in which ministries were in charge of environmental matters for the same resource that they exploited for productive purposes, making them both 'judge' and 'interested party' of the same activity."[18] The sugar industry was responsible for some 47 per cent of the polluting outflow load on coastal ecosystems. But that industry also pioneered recycling systems, using residues for energy production and an almost closed-system production design.

A decade later, the Institutes of Zoology and Botany finally merged into the present Institute of Ecology and Systematics. This group has led the way in developing programmes in biodiversity, protected areas, and protection of the coastlines and forests.

One issue that has not been fully resolved is nuclear power. To a country dependent on fuel imports, a nuclear power plant seemed very attractive. Soviet technical and economic help encouraged Cuba to begin the construction of a nuclear power station at Juraguá, near Cienfuegos. Misgivings arose: could Cuba really be safe in the event of a disastrous meltdown? In a small country, a major radioactive release would be even more devastating than in Russia. Would even normal operations without a catastrophic event poison the surroundings with radioactivity? Would the plant demand too much water? Could they be sure of finding a safe and secure home to put the deadly waste products? But while they pondered these questions, the Soviet Union collapsed, no further aid was forthcoming, and alternative forms of energy generation advanced. The half-plant still sits there, and the issue has been put on hold. Meanwhile, nuclear engineers have been encouraged to study alternative energy.

Cuba has, in practice, renounced the electro-nuclear path, partly because alternatives have appeared in the short run based on national oil and gas. "But doubtless," says the engineer José Luis Garica, "in strategic terms, we cannot rule out the possibility that at some time we might opt for electro-nuclear energy."[19] In September 2004, the principal turbine at the Juraguá plant was removed to replace a damaged turbine at the Guiteras thermoelectric plant. As in other fields, Cubans take a very broad view of the environment. The conception of an ecological pathway of development is emerging from the perspectives of conservation of natural areas, agriculture, public health, urban planning, alternative energy, clean production and waste disposal, community participation, environmental education, and issues involving different sectors of society, particularly vulnerable habitats. Workplace and neighbourhood pollution problems are included within the same framework.

The National Environment Plan of 1995[20] integrates a vast array of problems and proposals and is executed by the coordinated efforts of government agencies, NGOs and community participation.

AGRICULTURE

One of the outstanding achievements of Cuba's advance toward ecological development is the acceptance of agroecology as a national strategy. Agricultural development was at first dominated by the high-tech "Green Revolution" perspective of the international development community. But soon, Cubans in many institutions began a critical reevaluation looking at the economic structure of agriculture, geography of production, farm organization, pest management, soil fertility and mechanization.

This came about through the convergence of several different initiatives. Within agriculture, people such as Nilda Perez, Luis Ovies and Tenelfe Perez in plant protection, Miriam Fernandez in entomology, Magda Montes in citrus research, Rafael Martinez Viera and Antonio Castañeiras in the Alexander Humboldt Institute for Fundamental Research in Tropical Agriculture, and Ricardo Herrera in the Institute of Ecology and Systematics carried out projects in polyculture, soil microbiology and biological pest control.[21] Ecologists began to speak up against the pesticide treadmill.

There was fierce debate about pesticides and the ecological

pathway. The traditional progressivist viewpoint of European socialism was that there is an inevitable progression from "backward" to "modern." Capitalism inhibited the full development of the "modern" and monopolized its benefits while unloading the costs onto the workers and peasants. Therefore, the task of a liberated country was to proceed as quickly as possible along that pathway of "progress," avoiding the barriers inherent under capitalist governance. Features of agricultural modernization included the transition from labour-intensive to capital-intensive, from small-scale to the economies of large-scale, from the patchwork heterogeneity of peasant production to the rationalized homogeneity of agribusiness and specialized state farms, from subjection to nature to the conquest of nature, from superstition to scientific knowledge. Advocates of this approach saw themselves as rigorous materialists and mocked the ecological viewpoint as "idealist," sentimental nostalgia for some golden past that never really existed.

We fought back with the argument that it was the height of idealism to expect that we could pass resolutions about production and have nature obey. We proposed that development was a branching process in which technical choices were not socially neutral, and that each kind of society had to find its own pattern of relating to the rest of nature. Accumulating experience was showing that agroecology was productive, economical and safer than chemical means.[22] In particular, we argued that beyond the dichotomy of labour-intensive versus capital-intensive was a knowledge- and thought-intensive agriculture. Instead of mobilizing vast amounts of energy to move large masses of materials, we sought the design of systems that were as self-operating as possible. Mechanization was sometimes very important but at other times destructive of the soil, inefficient in very wet soils, too expensive, and a constraint on other agronomic practices. A combination of tractors and animal traction according to circumstances seemed a better choice.

Instead of having to decide between large-scale industrial type production and a "small is beautiful" approach a priori, we saw the scale of agriculture as dependent on natural and social conditions, with the units of planning embracing many units of production. Different scales of farming were to be adjusted to the watershed, climactic zones and topography, population density, distribution of available resources, and the mobility of pests and their enemies.

The random patchwork of peasant agriculture, constrained by land tenure, and the harsh destructive landscapes of industrial farming would both be replaced by a planned mosaic of land uses in which each patch contributes its own products but also assists the production of other patches: forests give lumber, fuel, fruit, nuts and honey but also regulate the flow of water, modulate the climate to a distance of about ten times the height of the trees, create a special microclimate downwind from the edge, offer shade for livestock and the workers, and provide a home to the natural enemies of pests and the pollinators of crops. There would no longer be specialized farms producing only one thing. Mixed enterprises would allow for recycling, a more diverse diet for the farmers, and a hedge against climactic surprises. It would have a more uniform demand for labour throughout the year. One example is the "El Carmen" UBPC, a new cooperative in Ciego de Avila. It has assumed national leadership in designing the transition from the previous conventional citrus monoculture to mixed production of fruit, annual crops and livestock products.

The arrogant presumption of the conquest of nature had to be replaced with a strategy of nudging nature here and there while respecting its autonomy and complexity.

Traditional knowledge could not be dismissed as

superstition but must be understood as a pattern of insights and blindness, just like modern science. Our task was to look at both of them critically in order to integrate the detailed, particular and nuanced peasant knowledge with the more general and comparative but abstracted knowledge of agricultural science, an integration that depended on scientists and farmers meeting as equals in a common enterprise. This is made easier by the fact that so many agricultural scientists come from peasant families. More recently, solidarity groups from New Zealand and Pro Naturaleza, an NGO organized by the Antonio Nuñez Jimenez Foundation, are spreading the Australian system of permaculture in Cuba.

The Special Period, with critical shortages of fuel, chemicals and feed, revealed the fragility of high-tech agriculture and encouraged the adoption of ecological agriculture. But it also reduced the capacity to carry out measures already adopted. Environmental inspections lapsed for lack of monitoring supplies and fuel to get to the sites to be inspected. Tough, thorny weed trees invaded fields abandoned for lack of tractors. Badly polluting buses were kept in service for lack of spare parts or replacements. Economic urgencies encouraged ignoring some protective regulations. We had the paradoxical situation that environmental conditions worsened while environmental consciousness deepened. When sound measures were introduced, some producers were convinced of their value only as an emergency measure. Our task became to convert these ecologists by necessity into ecologists by conviction before the emergency ended and they could resume the peaceful destruction of Cuba. This conversion is being carried out by education at all levels, the training of ecologically oriented agronomists, and ongoing debate.

Meanwhile in the 1970s and 1980s the Ministry of Defence, developed a new doctrine of defence that assumed the possibility of Cuba being partly occupied by a hostile power. The Cuban response would be a war waged by all of its citizens. But this required local self-sufficiency in the absence of central organization and exchange. A civil defence manual from that time had chapters devoted to first aid, herbal medicine, organization of schools, and food production. This led to military experiments in low-input agriculture. And in 1987 Raul Castro called for the widespread introduction of *organopónicos*, raised beds of enriched and composted soils where crops could be grown in small areas with no dependence on outside resources. The first pilot *organopónico* in Havana on Fifth Avenue and 44th street in Playa was organized by the armed forces and is still a showcase of urban agriculture. Today agriculture is evolving in the direction of agronomically and socially sustainable diverse production that emphasizes combining rural, suburban and urban farming; diversification; and biological and natural pest control.

COMBINING RURAL, SUBURBAN, AND URBAN FARMING

Urban agriculture now takes place on some 30,000 hectares producing more than 3 million tons of fresh vegetables per year for 11 million people. Like most Cuban programmes, it serves multiple purposes: It provides abundant, diverse fresh vegetables throughout the year for consumers. This has transformed the Cuban diet in the communities, schools and workplaces and encouraged the spread of vegetarian restaurants. It lowers the costs of transportation and storage by selling directly to consumers. It provides employment for some 300,000 people at a time when capital is not available to invest in more industrial employment. This comes to about ten people per hectare, a very labour-intensive system that would be regarded as highly inefficient in the United States, though each worker is producing ample vegetables to feed 36 people. In the context of the unemployment that appeared with the Special Period, it is socially efficient. Urban agriculture increases the green area of cities, detoxifies the air, and provides foci of neighbourhood social integration.

DIVERSIFICATION

Geographic diversification is a protection against regional disasters such as hurricanes, which may have zones of destruction 200 miles across. More locally, instead of having large monoculture farms, enterprises are converting to mixed production of fruits, vegetables, basic grains, livestock and fish. This results in a mosaic land use that makes better use of the topography and microclimates and permits recycling within the farm. Each patch of the mosaic has its own products but also contributes to the whole. As mentioned above, woodlands provide a wide range of forest products and ecological services. Pastures support livestock for meat and dairy products, manure for composting, combat erosion, and serve as nectar sources for honey and beneficial wasps. Oxen are integrated with tractors in a complex traction strategy, and horses and other animals are helpful in weed control. Diversification is a hedge against natural disasters that affect particular crops, and it more uniformly spreads the need for labour and assures local diversity of food supplies. By combining constant evaluation and decision making with the hard physical labour or farming, it also raises the technical level of agricultural labour for a rural population too educated to aspire to a life limited to wielding a machete. Composting, crop rotation, the use of nitrogen-fixing bacteria, fungi that mobilize potassium, phosphorus and other minerals, as well as the cultivation of earthworms maintain soil fertility.

BIOLOGICAL AND NATURAL PEST CONTROL

These methods are proving more effective than chemical control, more economical, and protective of people's health and the environment. Here's just one example: sweet potatoes grown using integrated pest management yielded 8.9 metric tons per hectare at a net per hectare value of $904.70 compared to 7.8 tons worth $818.60 per hectare for sweet potatoes grown with conventional methods using pesticides. Ecological protection makes use of polyculture and the spatial arrangement of crops, rotation, encouragement of predators, introduction of parasitic wasps and fungi, and, finally, biological products such as neem. All urban agriculture is now organic, and much of the rest of Cuban agriculture is advancing in that direction. The Antonio Nunéz Jimenez Foundation for Nature and Humanity is a major NGO in Cuba. It has developed strategic documents for urban sustainability and has been the leading group in the promotion of permaculture.[23]

The ecologists eventually won the struggle over developmentalism. It took a long time, and the debate was fierce at times, but it had a very different flavour from similar debates in the United States. All parties were looking for ways of meeting the needs of society so that the disagreements were just disagreements, not surrogates for clashing interests. Nobody was pushing pesticides or mechanization to make profit. And agroecology proponents were not demonized as "Luddites," or worse.

CUBAN SOCIALISM AND THE ENVIRONMENT

We can now return to the original question: how did Cuba do it? At the most abstract level, the short answer is socialism. That is, socialist social arrangements and ideological priorities made ecological development an almost "natural" correlate of the economic and social development and of the commitment to improving the quality of life as the primary goal of development. But an abstraction does not mobilize resources or change minds. Change occurs through the actions of particular people, through the decisions they make. And decisions are made in response to the questions that are posed, the social setting in which answers are sought, the tools available for providing answers, and the criteria for judging whether a solution is satisfactory. "Natural" only means that it was easier to make good decisions in the long run and that when poor decisions are made or institutional arrangements obstruct ecological rationality, there is a collective incentive to correct the errors.

The logic of decision-making under Cuban socialism starts with the priority given to human need. Therefore, questions about the environment arise more or less independently in areas such as urban development, health, agriculture, defence, conservation, and economics. There are no externalities—environmental harm that can be thrust upon the society as a whole and responsibility denied.

In each of these spheres, law reinforces the general ideological commitment. If one sphere lags, the others advance, and the convergence of ecological imperatives from different sources creates the direction of movement. The feedback mechanisms in the society favour ecology. Each success encourages further extension of the ecological commitment by showing that it is possible to develop in a way that departs from the conventional developmentalist wisdom. This is a positive feedback. When ecological rationality is subordinated to expediency and destructive decisions are made, these are visible, and there is a collective incentive to correct the error. A case in point is the stone causeway built from the Cuban coast to the tourist centre on Cayo Coco. Ecologists had warned that this would disrupt the hydrology of the area and harm the mangroves. But economic urgency prevailed. The causeway was built in spite of the warnings, and the mangroves began to die. But when this was observed, sections of the causeway were removed and replaced by bridge spans to permit the flow of the water.

In contrast, in capitalist society, each victory restricting the free destruction of our biosphere by business intensifies corporate resistance with an urgency to defend not only their profits but also their property rights. Thus there is a powerful anti-ecological negative feedback expressed as "backlash," amidst claims that environmentalists are "going too far." And environmental victories do not necessarily encourage further struggle.[24] When they don't unleash an environmental backlash, they are often co-opted as examples of "partnership." Of course there are also positive feedbacks at work under capitalism, which are important in movement building.

Socialism made ecological choices more likely. In spite of the incentives and commitments to an ecological pathway, Cubans could have decided otherwise. In fact, they did so at the beginning when in the absence of ecological consciousness, the urgency to meet the needs of the people led to harmful decisions. But when the first, Green Revolution developmentalist approach turned out to be destructive of productive capacity and poisoned people and nature, this was sufficient reason to reexamine the strategy. There were no greedy institutions committed to defending the harmful course with lobbyists, public relations firms, lawyers and hired witnesses. It meant that Cuban scientific and political leadership, which is strongly committed to a broad, dynamic and integral approach, was able to recognize the origins of the different developmental strategies in the world political economy and the implications of alternative choices. It meant that there were scientists prepared to argue the case for ecological development, receptive ears in the leadership and public to receive the arguments sympathetically, and a logic of decision-making that made an ecological pathway of development along with equity and collectivity an essential part of Cuban socialism. That's how they are doing it.

Postface
A View from the Trough

We are now living in the trough between two great waves of revolutionary struggle. The first wave began in 1848 and ended in defeat with the fall of Moscow. The second wave is still incubating. The collapse of the Soviet Union and the East European bloc was a tragic defeat for all of us, not because these regimes were models of the society we want but because they and the political movements influenced by them were the focus of the first world-wide challenge to capitalist power, capitalist exploitation, capitalist morality, and culture. Only in the context of that challenge did corporations agree to bargain with labour, the U.N. recognized the human rights to employment, food, education and culture, self-determination, women's equality, the illegitimacy of racism, and the rights of nations to autonomous development. After that collapse the New World Order rushes to deny those rights in theory as it always had in practice. It is no longer necessary to pretend respect for workers, and we see a new age of meanness in policy and in ideology. Lenin's anti-imperialist stance gave voice and form to the colonial liberation struggles that came almost to completion only after the Second World War. Almost, because Puerto Rico remains a colony, a residue of the old colonialism even as the new cycle of colonialism begins. The new colonialism is imposed mostly through economic means but also through armed interventions.

After the collapse we see the rush toward recolonization, as competition among the capitalists is released from the restraints of their common front against "communism" and they scramble for a new division of the world. Recolonization carries with it a revival of racism internationally and within our country. The upsurge in racism is not understandable as a psychological epidemic or the surfacing of long repressed human nature. It is a necessary correlate of the reconquest and redivision of the world. At home the open Nazis like John Metzger run interference for the David Dukes who run interference for the Pat Buchanans who run interference for the mainstream racists of moderation.

The socialist call for the full equality of women was never achieved either in the socialist countries or in the revolutionary parties. But women did win levels of participation not seen before; the socialist movement was a seedbed for left feminism, and the almost invisible accomplishments of the left show up again starkly as we see what happens after they are overthrown. For instance, Ingushetia, the autonomous region of Russia next door to Chechnya, now free of Soviet rule, has won the freedom to decriminalize polygamy and the sale of women in marriage. So in the new exuberant aggressiveness of world capitalism we see what communists and their allies had held at bay.

Finally, it was a massive defeat because almost all the socialist and communist parties have abandoned the struggle, their members scattered, some for the spoils of partnership, some to seek minor ameliorations of oppression under the slogans of "realism," some in anguish because history did them dirty, leaving a residue of despair and confusion even among many who struggled for decades with courage, imagination, and sacrifice for a better world that is not yet.

In a trough we can't see very far or identify what is immediately ahead. In any case Marxists have been notoriously unsuccessful in estimating the time it would take for major turning points. Marx expected a European revolution in the last century; but in 1916 Lenin wondered if he would see a revolution in Russia in his lifetime. Even in the bottom of the trough we can understand the sea and navigate by the stars.

I expect a new wave of upsurge because capitalism is not only what it has always been, but is more so now without the restraints imposed even by flawed non-capitalist states. It is less able and less willing to confront the challenges of environmental deterioration, new and resurgent diseases, chronic unemployment even during times of prosperity, the volatility of a business cycle mediated by financial instruments several steps removed from production, and a growing gap between rich and poor.

With the renewed intensity of international competition it has revealed a franker, more cynical viciousness in the recolonization of the third world and in waging the global class struggle against all workers. It has celebrated the spread of democracy while inventing new techniques for commodifying elections and thwarting real democracy. Its technological ingenuity makes possible the deeper penetration of the commodity relation into all corners of our lives, producing immense profit but also growing uneasiness.

The United States is fast becoming a second rate economic power while remaining a first rate military, police, and public relations power. This creates a specially dangerous situation, since it cannot but use its superior repressive power to strive for economic domination.

All of this makes it necessary for people to engage in struggle one way or another. Class struggle in fact does intensify by the bosses through down-sizing, speed-up, contracting out, two-tiered wage structures, union busting, take-backs, and relocation; by workers, often in individualistic ways. Illegal immigration, the massive theft from U.S. businesses by their under-paid and alienated employees, and robbery and street crime by the unemployed are but three modes of redistribution undertaken by the subjects of capitalism seeking to offset the growing concentration of wealth and callousness in the use of its power.

Necessity alone is no guarantee of success in confronting those necessities. However, people do perceive the intensifying problems through which the reality of our lives contradicts the pretensions of the society, and what is no longer automatically accepted as what must be.

I think a second revolutionary upsurge will come because I see the first signs of it: there is a disillusionment with the triumphalism of global capitalism even in Poland and Russia as people begin to see that all the lies they were fed, about capitalism by their censored media turned out to be true. French workers are asking whether the new Europe will be a new Europe only for the exploiters. There is a blooming of local peoples' organizations all over the world struggling for environmental justice, equality, health, and education. A recent directory lists 700-800 environmental organizations of people of color in the United States alone.

In the many conferences I attend I see that Cold War redbaiting has been losing much of its power to terrify as people raise issues which would have been forbidden ten years ago because they implicitly clash with the norms of the market. There is rage in Peoria today at the Caterpillar plant, and rage in the human service professions at the callousness with which the new stinginess condemns education, health care, social service, and science to frustrating inadequacy. There are not quite winds but at least fresh breezes in the U.S. labour movement. Even in universities, which were to some extent buffered against the worst excesses of the marketplace, business considerations increasingly overwhelm educational and intellectual concerns.

And it is at least noticed. There is awareness of the increases in the numbers and remuneration of administrators compared to faculty and staff, increased reliance on part time and temporary teachers, increased teaching loads, and repeated efforts to undermine job security that express a creeping proletarianization of what were once "free professions." And there is an increase in solidarity with Cuba not only because it is a target of aggression but also because it is an example of hope that things can be done differently.

But these glimmerings are still only incipient. We are at the cusp of a real historical discontinuity. We can see only hazily across that boundary. We cannot predict outcomes or events but only identify the contradictions that are developing, the problems that have to be confronted again and again until some resolution is reached. We do not know what roles self conscious Marxists or the religious left will play in the new movements, how much it will remember from the past or have to learn anew. We do not know how movements for specific causes or on behalf of particular oppressed constituencies will relate to political movements that embrace all the liberating struggles. We do not know what relationships will arise between the political movements and their intellectual organs that best combine the needs for responsible engagement, the mass mobilization of a collective intelligence, and the relative autonomy of innovative inquiry. We do not know what new holidays will bring us to the streets or the beat of the new songs.

But without knowing the shape of the future we can still identify some of our tasks:

1. Assist in the revival and growing clarity of popular struggles, helping the new movements to broaden their vision, to understand the context of their immediate situations and the lessons of past struggles that they can draw on. The long view is vital in sustaining the short term and local struggles in

the face of countless disappointments and frustrations, anticipating the pressures of our adversaries to divide and co-opt, and in discovering the common ground between different struggles for justice when they seem to conflict because each asks too little.

2. Combat despair. As against the retreat into nationalism we reaffirm our internationalism, maintain ties among revolutionaries across borders, pool our experiences and ideas, and work for joint strategies. We uphold special solidarities with Cuba and Puerto Rico. Puerto Rico because as one of the last of the old colonies it may spearhead the resistance to the recolonization of the Third World. Cuba because it is the one society which has retained a revolutionary commitment although having to make concessions to global capitalism. Even as it retreats in some spheres of life it leads the world in others, in the vision of an ecological society and the development of socialist democracy.

As against individualism we learn from left feminism to examine the much ignored areas of the personal in society, explaining how our individual miseries are not ours alone, how the commonly marketed solutions to personal fulfillment fail to address the roots of the prevailing miseries and wasted talents. We at least can raise and explore the contradiction between the personal responsibility that is a prerequisite for tolerable and even joyful lives despite an intolerable society and the social causation that places our own situations in perspective and identifies the targets of our public struggle.

As against the ideologies of despair, we challenge the theories that misuse linguistic analysis or the mathematics of chaos to proclaim the impossibility of understanding the world. We challenge the misuses of the real uniqueness of every place and every person to deny the possibility of general theories about anything. We challenge the misuses of the awareness of science as a social product to argue that all theories are equally invalid and the categories of science just objects of discourse. We challenge the fatalistic arguments of the genetic determination of inequality or meanness or aggression. We do all this by showing the political instrumentality of such ideas and locating them in the history of ideology. We do this also by confronting them with a dialectical understanding of the relations between chance and determinacy, the general and the particular, the partisanship and objectivity of science, the internal and external in causation. Dialectical materialism is a practical, urgent weapon in building a movement by providing a better way to understand the world and ourselves in it.

3. Faced with the dismissal of Marxism as obsolete even by some progressives, we reaffirm a militant Marxism. Rather than shrinking it down to merely a humane economics in order to gain respectability we broaden the scope of its engagement to confront all the ideologies of aggressive capital in all aspects of our existence. Only an honest, creative, and self-critical Marxism can survive to play a vital role in the coming struggles.

a) As a matter of theoretical coherence, practical necessity, and intellectual integrity we have to examine the history of our movement and understand the defeat it suffered. We have to examine the world and national conditions that besieged socialism and made it vulnerable to its enemies. We have to examine the ins and outs of Rosa Luxemburg's paradox, the contradictions that arise from building the new with the materials of the old. We have to look at the beliefs and practices that undermined socialism's development from within. This includes an honest appraisal not only of our errors but also of our crimes. Errors are misjudgments in the service of our agreed-upon programme, unnecessary compromises or pompous refusals to compromise, faulty estimates of our progress and the enemy's weakness, passive acceptance of capitalist ways of doing things in the hope that they could be domesticated to socialist ends. Crimes are violations of socialist democracy, socialist legality, revolutionary humaneness, and that fierce honesty which is basic to the commitment to liberate and mobilize the collective intelligence of all the oppressed. Crimes are the debasement of Marxism to apologetics, the use of force to settle disagreements within the revolution, the covering up of corruption.

In dealing with the crimes committed in the name of communism we cannot opt for the easy solution and simply expel Pol Pot, and Beria, and so on from our ranks, deciding that they weren't good communists after all and surely had nothing in common with us. That way we shed responsibility, feel good about ourselves and learn nothing. There is, of course, truth in the claim that these horrendous episodes are not communism but distortions of communism just as we might say that corn smut is not corn but a disease of corn. But there is also the other side of it: corn smut is a disease of corn, not of tomatoes or orange trees. It could only take hold in a vulnerable substrate, and our task must be to understand how to eliminate that vulnerability before we can expect people to take another chance. Thus we affirm the dual propositions: the triumph of centralism over democracy, the suppression of dissent among revolutionaries, the turning of labour from a rehabiliation process into a cruel punishment, all of these are not communism but *distortions* of communism; but also they are distortions of *communism*.

b) Openness to new ideas. Just as Marxism acknowledges its debts to English political economy, German philosophy and French socialism so it must also welcome the insights of feminism, national liberation and anti-racist struggles, and ecology. These are not alien to Marxism. Militants in all of these struggles include Marxists, people who were familiar with Marxism, people who passed through Marxist parties. Marxist thought has left its imprint on all of them. Marxist thought was also familiar with and influenced by feminism, nationalist thought, ecological ideas. All of these movements have overlapped in ideas and people and will continue to do so.

c) Openness to new phenomena, to changes in our society and in the ways people confront that society, to new patterns of consciousness. Alertness to the new must be rooted in our understanding of the past so that we don't go around proclaiming new paradigms capriciously or declaring ideas obsolete because they were written in the last century or because they seem to go against the trend of the moment. The decline of employment in the traditional basic industries of steel, coal and autos does not mean the end of the working class in a country where the overwhelming majority of the people still live by selling their labour power. The diversification of forms of exploitation in Africa and Latin America beyond simple extraction of minerals and agricultural produce does not mean the end of imperialism. The failures of overly centralized planning of the economy do not make the market the only alternative to stagnation.

But without a self-conscious rush to be revising Marxism we do have to note the new. The massive and pervasive threats to the integrity of our biosphere, new patterns of linking local hierarchies to transnational corporations, changes in the technical structure of industry and finance, new kinds of grass roots organization popping up all over, a new willingness of people to go beyond challenge to local abuses.

We have to examine and invent new forms of struggle, all aimed basically at changing consciousness and building solidarity even when we are small and seemingly helpless. Revolutionary politics is not limited to storming the winter palaces. Any action that pushes back the boundaries of the permissible, that legitimizes thinking and questions the unquestionable, that strengthens our own capacity to analyze and organize and that tightens the ties that unite us for the long haul, that invents ways of broadening participation and that undermines the crippling burdens of racism and sexism and homophobia and hierarchical posturing within our own movements, is revolutionary practice.

Nor do I put down what is derisively called "preaching to the converted." We, the "converted," need lots of "preaching," lots of analysis, education, encouragement.

From the bottom of the trough we need to see the present moment in perspective, to know that it isn't over, that even when exuberant capitalism wins big victories these do not solve its problems. The problems return even more sharply. Therefore the struggle will surge again, and we will add new pages to our songbook. I expect to see you there.

Notes

WHOSE SCIENTIFIC METHOD?

[1] A.R. Omran, "The epidemiological transition. A theory of the epidemiology of population change," *Milibank Memorial Fund Quarterly*, vol. 49, no. 4, 1971, pp. 509–38 and Richards Levins, T. Awerbach and U. Brinkmann, "The Emergence of New Diseases," *American Science*, no. 82, 1994, pp. 52–60.
[2] R. Levins, *Evolution in Changing Environments*, Princeton: Princeton University Press, 1968.
[3] R.C. Lewontin and R. Levins, "Schmalhausen's Law," *Capitalism, Nature, Socialism*, vol. 11, no. 4, 2000, pp. 103–08.
[4] E. Goodman, B.C. Amrick, M.O. Rezendes, S. Levine, J. Kagan, W.H. Rogers, and A.R. Tarlov, "Adolescents' Understanding of Social Class: A Comparison of White Upper Middle Class and Working Class Youth," *Journal of Adolescent Health*, vol. 27, no. 2, 2000, pp. 80–83.
[5] M.R. Reich, *Toxic Politics. Responding to Chemical Disasters*, Ithaca: Cornell University Press, 1991.
[6] *Birdscope*, vol. 15, no. 2, Spring 2001.
[7] *Granma*, March 28, 2002.
[8] R. Levins, "Toward a Population Biology, Still," *Evolution of Population Biology*, Rama Singh and Marcy Uyenoyama (eds.), Cambridge: Cambridge University Press, 2004.

CONVERGENCE OF DEVELOPMENT GOALS

[1] Agustin Lage, *La economy de conocimiento y el socialismo, II. Reflexiones a partir del proyecto de desarrollo territorial en Yaguajay. Cuba Socialista* #33, 2005.

THE SCIENCE OF DHARMA AND THE DHARMA OF SCIENCE

1 For the purposes of this essay, I am using Meera Nanda, *Breaking the Spell of Dharma and other essays*, New Delhi: Three Essays Press, 2002. All citations in this chapter are from this book. In addition, for a fuller statement, please see Meera Nanda, *Prophets Facing Backward: Postmodern Critiques of Science and Hindu Nationalism in India*, New Brunswick, NJ: Rutgers University Press, 2003.

2 On "anti-Americanism": Revolutionaries are not anti-American, but imperialist and internationalist. In my own experience in Latin America and Vietnam, I have never encountered anti-Americanism among revolutionaries. Rather, I experienced it among members of the ruling class, who accepted the domination of the United States and expressed their resentment in chauvinism and cultural nationalism.

IS CAPITALISM A DISEASE?

1 Robert Fogel, *Time on the Cross: The Economics of Negro Slavery*, Boston: Little, Brown and Company, 1974.

CUBA'S FIDELITY TO THE EARTH

1 The following references introduce each of the areas of Cuban achievement not discussed in this paper. Jerry M. Spiegel and Annalee Yassi, "Lessons from the Margins of Globalization: Appreciating the Cuban Health Paradox," *Journal of Public Health Policy*, vol. 25, no. 1, 2004, pp. 85–110; Anuario Estadistico, Ministerio de Salud Publica de la Republica de Cuba, Candido Lopez Pardo, Miguel Marquez and Francisco Rojas Ochoa, "Desarrollo Humano y Equidad en America Latina y el Caribe," XXV Internacional Congreso of the Latin American Studies Association, October 2004; Francisco Rojas Ochoa and Candido Lopez Pardo, "Desarrollo Humano y Salud en Ame´rica Latina y el Caribe," *Revista Cubana de Salud Publica*, vol. 29, no. 1, 2003, pp. 8–17; United Nations Development Programme, "Report on Human Development, 2004," published online at http://www.undp.org; Mario Coyula and Jill Hamburg, "Understanding Slums: the Case of Havana, Cuba," David Rockerfeller Centre for Latin American Studies, no. 4, 2004–05; Maria Caridad Cruz and Roberto Sanchez Medina, *Agriculture in the City: A Key to Sustainability in Havana, Cuba*, Kingston, Jamaica: Ian Randle Publishers, 2003, Miren Uriarte, *Cuba Social Policy at the Crossroads: Maintaining Priorities, Transforming Practice*, Boston: Oxfam America, 2002.

2 Minor Sinclair and Martha Thompson, *Cuba Going Against the Grain: Agricultural Crisis and Transformation*, Boston: Oxfam America, 2001, p. 56.

[3] For the purposes of this study, socialism refers to a society in which the associated producers (past, present and future workers) own most of the means of production locally or nationally, in cooperative or state enterprises, and make decisions about the society through combined participatory and representative democracy (both state and non-state). Production is decided on the basis of judgments about human needs, distribution is according to work and need, and labour power is not a marketable commodity. Within this framework many kinds of organization and ways of working have been and are being tested.

[4] Complexity has become a fashionable buzz word in discussions about science, a recognition that fragmented, reductionist, and ahistorical science has caused major disasters and left us unprepared for the resurgence of infectious disease, for antibiotic resistance in bacteria and pesticide resistance in agriculture, for flooding as a result of flood control engineering, for the emergence of hospitals as foci of infection, food aid leading to hunger, and economic growth giving rise to new forms of poverty. The rediscovery of complexity emphasizes uncertainty, non-linearity, connectedness, the interaction of chance and determinism, and the near orderliness of chaos. In one sense it is a groping toward dialectics without acknowledging the Marxist tradition. See Richard Levins, "Dialectics and Systems Theory," *Science and Society*, vol. 62, no. 3, 1998, p. 375; and Richard Levins, "The Butterfly ex Machina," in Rama S. Singh, Costas B. Krimbas, Diane B. Paul and John Beatty (eds.), *Thinking About Evolution, Vol. II: Historical, Philosophical, and Political Perspectives*, Cambridge: Cambridge University Press, 2001, p. 606.

[5] Josefina Toledo Benedit, *La Ciencia y la Técnica en José Martí*, Havana: Editorial Ciencia y Técnica, 1994, p. 209.

[6] The Cuban Speleological Society was organized by teenage boys from Havana who loved exploring the countryside. They began as a Boy Scout troop but broke away when they decided that the Boy Scouts were a conservative, militaristic organization dominated by the United States. The geographer Antonio Nuñez Jimenez was a member and introduced Fidel Castro to cave exploration in the early years. The Society continued as a small ecology-minded NGO.

[7] Silvia Martinez Puentes, *Cuba Mas Allá de los Sueños*, Havana: Editorial Jose Marti, 2003, p. 112.

[8] Martinez Puentes, *Cuba Mas Allá de los Sueños*, p. 128.

[9] Fidel Castro, speech at the closing of the Congress of Pedagogy, February 2, 2003.

[10] J.D. Bernal, *The Social Function of Science*, London: Routledge, 1939.

[11] Jorge Aldoregía Valdes-Brito, Pablo Resik Habib, and Hector Rodríguez Báster, "Organización y Administración de la Investigación," *Revista Cubana de Administración de Salud*, no. 11, 1985, pp. 305–16.

[12] Carlos Borroto, discussion in Fidel Castro Diaz-Balart, *Cuba. Amanecer del*

Tercer Milenio, Havana: Editorial Debate, 2002, p. 257.

[13] CITMA (Ministerio de Ciencia, Tecnología y Medio Ambiente), Taller "Medio Ambiente y Desarrollo: Consulta Nacional Rio.5," Havana: CITMA, 1997, p. 74.

[14] Ibid, p. 22.

[15] The Second Biennial International Seminar on the Philosophical, Epistemological and Methodological Implications of Complexity Theory and Parallel Workshop on Complex Biological Systems, Havana, January 7–10, 2004.

[16] Agostín Lage, discussion in Fidel Castro Diaz-Balart, op. cit, p. 145.

[17] Gina Rey, "Cuba, Integral Development, and the Environment," Presentation at Dalhousie University, 1989.

[18] CITMA, National Environment Strategy, 1997.

[19] José Luis Garcia, discussion, in Fidel Castro Diaz-Balart, op. cit., p. 306.

[20] CITMA, Programa Nacional de Medio Ambiente y Desarrollo, 1995.

[21] The people mentioned in this chapter are an idiosyncratic sampling of names I happen to remember and people I have worked with. Others whose contributions are just as important have been omitted.

[22] Fernando Funes, Luis Garcia, Martin Bourque, Nilda Perez and Peter Rosset, *Sustainable Agriculture and Resistance: Transforming Food Production in Cuba*, Oakland: Food First Books, 2002, p. 307.

[23] Maria Caridad Cruz and Roberto Sanchez Medina, *Agriculture in the City*.

[24] A classic example is the case of Alar, a possible carcinogenic chemical used on apples. Environmentalists were successful in getting Alar removed from the American market in 1989 following a major exposé on the CBS News programme, 60 Minutes. But this turned into a major defeat for environmental activism after a food and chemical industry-sponsored public relations campaign successfully spun this legitimate victory into "environmental fearmongering," a label that has become a potent weapon against environmentalism in general.